ALLIED HEALTH CHEMISTRY

A Companion

Timothy Smith • Diane Vukovich
University of Akron   *University of Akron*

Prentice Hall

Upper Saddle River, NJ 07458

Acquisitions Editor: Mary Hornby
Production Editor: James Buckley
Cover Designer: Paul Gourhan
Production Supervisor: Barbara A. Murray

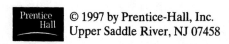 © 1997 by Prentice-Hall, Inc.
Upper Saddle River, NJ 07458

Printed in the United States of America

10 9 8 7 6 5

ISBN 0-13-470460-6

Prentice-Hall International (UK) Limited, London
Prentice-Hall of Australia Pty. Limited, Sydney
Prentice-Hall Canada, Inc., Toronto
Prentice-Hall Hispanoamericana, S.A., Mexico
Prentice-Hall of India Private Limited, New Delhi
Prentice-Hall of Japan, Inc., Tokyo
Editora Prentice-Hall do Brasil, Ltda., Rio de Janeiro

Contents

Chapter 1	Taming Chemistry	1
Chapter 2	Just a Little Math	6
Chapter 3	Push Button Thinking *Using Your Calculator*	28
Chapter 4	Rulers, Weights, and Measuring Cups *Measurement and the Metric System*	46
Chapter 5	It's the "One" Thing to Do *Unit Conversions*	58
Chapter 6	Bigger and Better Conversions	69
Chapter 7	Why? Why? XY *Simple Algebra*	77
Chapter 8	Light Headed With All These Heavy Thoughts *Density*	95
Chapter 9	When You're Hot *Temperature Conversions*	108
Chapter 10	Counting Calories *Heat and Energy*	116
Chapter 11	Practice Tests	123
Chapter 12	Furry Little Unit Conversions *Moles*	127
Chapter 13	Know Any Good Recipes? *Stoichiometry*	133
Chapter 14	A Lot of Hot Air *Gas Laws*	139
Chapter 15	How Strong is That Coffee? *Solution Concentrations*	150
Chapter 16	The Basics on a Sour Subject *Logarithms and pH*	157
Practice Problems		167

Study Skills Sections

Good Study Strategies	26
Get to Know Your Textbook	44
How to Take Notes in Class	56
Reading Your Textbook	67
Learning to Relax	92
Preparing for the Test	106
Taking the Test	114
Mastering the Multiple Choice Test	121

Acknowledgments

We are grateful to our colleagues and students who gave us valuable suggestions and constructive criticisms.

We would like to thank Aleksandra Tomich for encouraging us to start this project, our editor Mary Hornby for her persistence and enthusiasm, Christina Anderson for her illustrations and Jeffrey Swainhart for scanning them, and Rita Klein, Ann Kaylor and Mary Ann Albanese for their proof reading help. We would also like to thank the following people who reviewed the manuscript and made helpful suggestions for its improvement:

Jean Ann Lanier, Tarrant County Junior College

Joanne Lin, Houston Community College

Edward Alexander, San Diego Mesa College

Barbara Mowery, Thomas Nelson Community College

Linda Wilson, Middle Tennessee State University

Mary Nelson and Robert Nelson, Georgia Southern University

Tim thanks Jill Strahler, Kathy MacIntyre, Mary King, Amie Ellis-White, Dana Philage, Dawn Dearth, Bonnie Graham and Brady's Cafe, and Suzanne Holt and the Zephyr.

Diane wishes to give special thanks to her husband, Tom, for his love, support, and understanding through this project.

Finally we want to thank Diane's cat, Petunia, and Tim's cats, Yoshi and Tifi and Fey for helping us keep our sense of humor.

How to Use This Book

This book will not replace your textbook. However, like a patient friend, this handbook will hold your hand during the first weeks of class and help you get started on the right foot.

A chemistry class is not a mathematics class, but chemistry requires some knowledge of mathematical concepts. Because many students entering chemistry classes have a limited background in mathematics, they trip up in the first weeks of class and never catch up. Often students struggle so much with the mathematics that they do not have the time to learn the concepts of chemistry. Our hope is that this handbook will help you avoid that mathematical stumbling block.

How are the Chapters Arranged?

The first 10 chapters of this book discuss in detail all of the mathematical concepts and operations which you will need to be successful in your first chemistry class. These chapters will help you to succeed on your first exam and then build a solid foundation for success in the remainder of the term.

In Chapter 11 you will find two samples of multiple choice tests which cover the types of mathematical material usually found on chemistry tests early in the term. After studying the materials in Chapters 1-10 and before your first major exam, try these short tests as if you were taking them in class. The library is a good place to do this activity. Then you can check your answers against the answer key.

In Chapters 12 through 16 we cover mathematical and chemical concepts which you will encounter later in the term such as moles, stoichiometry, gas laws, solutions and pH.

The first 11 chapters get you rolling along; the last 5 chapters give you a little helping hand along the way.

What to Look for in the Chapters

Looking Ahead

Looking Ahead—These sections give you a preview of "coming attractions." They show how your current work is very similar to what you will be doing later in the class. You are not expected to completely understand what the terms in these sections mean. But you will see that the patterns for doing problems later in the class will be similar to the patterns you are learning in each chapter. These sections peek ahead and show you that your chemistry class will not become much more complex than it is at the beginning.

Important
Point

Important Point—These small sections point out special things to keep in mind while studying or working problems. They alert you to little tricks you can do to make solving a particular problem a little easier or traps to look out for when doing a certain type of problem.

Study Skills Sections—Chapters 2 through 10 (except for Chapter 6) end with special Study Skills Sections that provide helpful hints on preparing yourself for chemistry. These how-to's include effective note taking skills, better techniques for reading your textbook, foolproof methods for studying chemistry as well as suggestions on how to decrease tension while studying, reduce test anxiety and develop test taking strategies.

Practice problems—At the end of the book you will find several practice problems for each of the chapters (except for chapters 1, 4 and 6). Doing these problems will reinforce what you have learned. There is also an answer key, complete with problem setups and solutions.

CHAPTER 1

Taming Chemistry

There is a fable about a man who encountered a lion in the forest. At first he was terrified and started to run. But then he saw that the lion was limping with a large thorn in his paw. He slowly approached the lion and carefully pulled out the thorn. The lion was grateful. Years later the man was taken prisoner and condemned to be thrown to the lions. He was led into the Colosseum and the lion, released from his cage, bounded out ferociously. But when the lion saw the man, he stopped, and instead of attacking him, he licked him in the face. The lion was the one he had befriended years before. The lion's behavior was so unexpected that both the man and the lion were set free.

What does this little story have to do with chemistry? Well, to many people chemistry may look like a beast that will devour them or at least block their way to where they want to go. But just like the lion, with a little care, you can tame chemistry, and maybe years later chemistry will come to be your friend. We hope this book will help you tame chemistry.

You Are Smart Enough to Study Chemistry

Many students believe that someone has to be extraordinarily intelligent to succeed in chemistry. And you may worry that you are not smart enough. Don't be afraid. Many normal people have successfully studied chemistry. You *do* need to have a good grasp of basic mathematics and you *do* need to be organized in your study habits. This book will help you with both of these.

1

Many students think that some people are "chemistry types," and they are not one of them. They label themselves or let others label them and then they begin to believe the label. Although some may love chemistry so much that they will go on to become chemists, and you might label them as "chemistry types," the chemistry in your class (and even beyond!) is well within the grasp of people with normal intelligence.

Chemistry is Relevant to Your Life

Many students believe that the chemistry class they are required to take is just a hurdle they must jump to get to their major. They believe that chemistry is only for chemists and irrelevant to their lives. But what we hope you will begin to realize is that your whole life is chemistry. Whether it's the food you eat, the medications you take, the adhesive on the back of your stamps or the ink and paper of this book, everything around you involves chemistry.

When you know more chemistry, you will be better informed to make medical, health and nutrition decisions in your own life. We are all confronted with environmental crises such as acid rain and the depletion of the ozone layer. We as citizens need to make decisions about these issues and without at least a basic knowledge of chemistry, making intelligent decisions about such matters is almost impossible.

Chemistry Plays a Big Part in Your Career

Students don't realize when they are studying chemistry what a big part it will play in their profession. One of our students, a registered nurse who came back to school to get her bachelor of science in nursing, made a list of chemistry related activities she encountered in one of her weeks on the job:

- Set up 2% lidocaine solution in D5W (5% glucose in water).

- Patient in cardiac arrest given calcium chloride ($CaCl_2$) to strengthen the heart contractions.

- Same patient given sodium bicarbonate ($NaHCO_3$) to control blood pH. The patient's blood was acidic from respiratory acidosis.

- Hydrated a patient with 0.9% sodium chloride (NaCl) solution because this is the concentration of normal blood.

- A patient with heartburn was given the antacid Maalox (200 mg aluminum hydroxide, $Al(OH)_3$ and 200 mg magnesium hydroxide, $Mg(OH)_2$).

- Received a patient's blood electrolyte report (Na^+, K^+, Ca^{2+}, Cl^-, HCO_3^-, HPO_4^{2-}). This patient had edema because of electrolyte imbalance.

- Asthmatic patient given O_2 to increase the partial pressure in the patient's lungs.

- Oxygen causes combustion. Precautions must be taken when using O_2 for patients, i.e., no flames, petroleum products.

- Administered a diuretic to eliminate excess fluid from a patient who was in congestive heart failure. Potassium ions are also eliminated with the fluid and must be replaced to prevent the effects of low blood potassium.

- Patient given barium enema ($BaSO_4$) to X-ray lower G.I.

- Used 70% alcohol as a disinfectant.

- I.V. drip set at 20 drops per minute (40 drop = 1 mL).

- Icebag was put on a sprained ankle to prevent swelling. (The ice melting absorbs heat and cools the ankle.)

- Measured a patient's blood pressure with a sphygmomanometer to be 120/70 mm Hg.

- We have to know the compatibility of intravenous drugs to prevent precipitations which might cause an embolism.

- Ordered specific gravity test of a patient's urine to check kidney function.

She went on to write:

> What is Chemistry? It is nothing more than life itself. As a nurse I am confronted with the need for knowledge of chemistry whenever I give care to my patients. It is imperative that I understand the normal chemistry of the human body so I can intervene effectively when the abnormal occurs.

Your Chemistry Class is Not a Math Class

Many students think that a chemistry class is just another type of math class. There *is* mathematics in chemistry, but chemistry is primarily descriptive. Chemistry is a collection of observations of our everyday world along with explanations of why they occur. When you note that a loaf of bread does not come out right if you add too much or too little yeast or sugar, you are doing chemistry. Chemistry has much more in common with cooking than with mathematics. In fact cooking is chemistry.

There is not a great deal of algebra in your class and there are only a dozen or so algebraic equations. This book will work with you on the mathematics you do need.

Your Chemistry Class Will Give You a Chance to Learn about Yourself

Your chemistry class will not only begin teaching you about the chemical processes that occur in your body—which is certainly learning about yourself—but will also teach you something about your own psychology. How do you study for tests? How do you learn to relax when taking tests? By watching and learning about these things now, you will learn something more valuable than the chemistry. You will gain valuable self–knowledge that will help you in your advanced classes and the Board tests.

Don't be Affected by Others' Negative Attitudes Toward Chemistry

Has anyone said to you when he or she found out that you are taking chemistry, "You are a glutton for punishment," or "You're in for it now"? Many people have preconceived notions about chemistry. They might have had a bad experience in high school. Or they might just be repeating what they have heard other people say. Don't be swayed by others' negative attitudes. Form your own judgment from your own experience.

In your class, surround yourself with positive people. Avoid negative people who constantly complain about the lectures, the textbook, and the homework. These students often blame others for their lack of success and refuse to take responsibility for their own destiny. They don't like to feel alone, so they will find other students to convince to feel the same way. They need confirmation for their negativity. Don't fall into this trap! Find the students in the class who have the qualities that you want to have—the students who will help you and inspire you to do your best.

Approach Chemistry With Enthusiasm

Instead of seeing your class as a burden or a chore, look at it as an exciting opportunity. Approach chemistry with enthusiasm. If at first you don't feel enthusiastic—pretend! It often happens that if we act positive and motivated, our mind starts to believe it. Tell yourself that you're really excited and eager to go to lecture or to read your textbook. Tell yourself, "I'm bright and intelligent and can learn anything I choose to learn."

Be Curious

Be curious. A chemistry class is a great opportunity to begin to learn about little things in life you may have wondered about: Why do we sweat? What's in an aspirin tablet? Why do salt crystals form little cubes? What are the ingredients in shampoo? And even, why is the sky blue? Again, if you're not curious—fake it. Just think—after finishing your class you'll be able interpret the ingredients lists on labels.

Remember: Some Confusion is Natural

Everyone is always a little confused and disoriented when confronted with new ideas and concepts. If you know students in an advanced math or chemistry class, it may not seem like it, but they struggled in their beginning courses and continue to struggle as they progress. They were probably as bewildered as you sometimes feel.

Your Success in Chemistry is Directly Proportional to Your Feeling of Self-worth

When your feeling of self-worth increases, your success in chemistry and other courses increases. This is actually a relationship similar to ones you will examine later in your class (and which we will look at in Chapter 7). When one goes up, the other naturally goes up. A high feeling of self-worth means that you believe you can succeed and not only pass your chemistry class and go on to your program and graduate, but also make a difference in your profession, win the respect of your colleagues, and truly help the people you serve.

When you are doing homework problems and studying for tests, remember what your long-term goals are. Sometimes it's easy to forget them in the everyday grind. Keep reminding yourself. And remind yourself that you are an important person with important things to do, and your chemistry class is one step on the way.

CHAPTER 2

Just a Little Math

In this chapter and in the following chapters, we will look at the math you will need to understand chemistry. You'll be happy to learn that it is neither extensive nor extremely difficult. There are no calculus problems or quadratic equations. You do not have to plot complicated graphs. In fact the algebra you do need to know is fairly simple. Furthermore, once you set up a problem, you will do most of the computations on a calculator.

In this chapter we will review with you the mathematical concepts you need to know in order to look at answers to numerical problems and say, "Yes, this answer makes sense" or "that looks right." In the next chapter and in the rest of the book, we will show you how to do the calculations quickly and easily on a calculator. In Chapter 7 we will go a little more in depth into the algebra and equations you will need later in the term.

Which Way Do I Go?
Positive and Negative Numbers

In ancient times, our ancestors only used simple numbers like 1, 2 or 3. Maybe they had 3 sheep or they had 5 goats. Life was less complicated then. They did not have a symbol for "nothing" for which we now have the number "zero." Europe only got the number zero about 500 years ago from Arabic traders. Then during the Renaissance, merchants and traders also invented negative numbers to help keep track of their debts.

In mathematics we have names for these different groups of numbers. Plain old numbers like 1, 2, 3, etc., are called "natural numbers." They are called natural because they are the numbers people first used to count things. We can imagine owning 5 goats. But we can also imagine owning zero (0) goats. We can even imagine owning negative five (−5) goats, which would be the case if we owned no goats but owed 5 goats to our neighbor.

The positive numbers, the negative numbers, and zero are collectively called signed numbers. If the numbers are whole numbers, such as 5, 0, or –5, they are called integers.

You can relate positive (+) and negative (–) numbers to money you have and to money you owe. (Remember that is how negative numbers got their start.) For instance if you have $800 in the bank, you have +800 dollars or simply 800 dollars. (If a number has no sign in front of it, we assume that it is positive.) If you owe $900 on your credit card, you have a debt which you write as –900 dollars and read as "negative 900 dollars." In this case, if you add up the money you have and the money you owe, you can see that you are 100 dollars in the hole, or that you "have" –100 dollars.

The common way to portray signed numbers is with a number line:

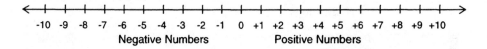

Zero is the center of the number line. The positive numbers go to the right: the negative numbers go to the left. Every positive number on the right has its opposite negative number on the left. A number that is farther to the left on the number line is smaller than a number on its right. For example, 3 is smaller than 8 so we can write

$$3 < 8$$

which we read as "3 is less than 8." If you have 3 dollars, of course, you have less money than if you have 8 dollars.

If you have an 8 dollar debt you are worse off than if you only owe 3 dollars. We can write this relationship as

$$-8 < -3$$

Notice that –8 is farther to the left on the number line than –3.

You can also relate signed numbers to the temperature scale. A thermometer is nothing but a vertical number line. Warmer temperatures, of course, are higher on the thermometer than colder temperatures. 20 degrees is warmer than, or greater than, 10 degrees.

$$20 > 10$$

But twenty degrees below zero (–20) is colder than, or less than, ten degrees below zero (–10).

$$-20 < -10$$

On a vertical number line, positive numbers go up and negative numbers go down. Positive five (+5) is five steps up from zero. Negative five (–5) is five steps down from zero.

Important Point

Remember to relate new concepts to something you already know. Most of you have been balancing a checking account and reading thermometers for years, so you already know a lot about signed numbers.

Where Will You See Signed Numbers?

Three places you will routinely encounter positive and negative numbers in your chemistry class are:

- charges on subatomic particles such as protons and electrons and also on charged atoms called ions.

- exponential and scientific notation.

- temperature scales as degrees above and below zero.

Adding Signed Numbers

It is simple to add two positive numbers because it is exactly the same kind of addition you did in grade school math. Two positive numbers add to make a bigger positive number, just like we see in everyday life. Two assets add together to make a bigger asset. If Joe has 7 dollars and Mary has 8 dollars, when they combine their wealth they will have 15 dollars:

$$7 + 8 = 15$$

We can show this addition on the number line as:

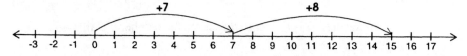

If you go 7 steps to the right of zero and then continue 8 more steps to the right, you wind up 15 steps to the right.

Two negative numbers add to make a more negative number, just as two debts add together to make an even bigger debt. Let's say Joe is $8 in debt and Mary is $7 in debt. When they get married they will "have" –15 dollars, that is they will be $15 in debt:

$$-8 + (-7) = -15$$

Notice the –7 is in parentheses. We do this to separate the addition sign from the negative sign on the 7. We can show this addition on the number line as:

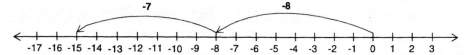

If you go 8 steps to the left of zero and then continue another 7 steps to the left, you wind up 15 steps to the left.

When you add a positive and a negative number, the sign on the answer depends on which number is larger. If Joe is seven dollars in debt and Mary has saved fifteen dollars, what do they have when they combine their resources? They end up (thanks to Mary) with 8 dollars in the bank:

$$-7 + 15 = 8$$

We can show this addition on the number line as:

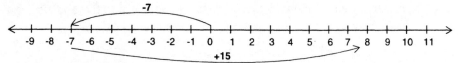

If you go 7 steps to the left of zero and then move 15 steps to the right, you wind up 8 steps to the right of zero.

When you add a positive number and a negative number, you subtract the smaller number from the larger and give the answer the sign of the larger number.

Let's look at a few more examples.

$$-6 + (-5) =$$

The signs of both numbers are negative, so add the numbers and make the sign negative.

$$-6 + (-5) = -11$$

Now try this example:

$$-10 + 13 =$$

The numbers have opposite signs, so subtract the smaller number from the larger number ($13 - 10 = 3$). Since the sign of the larger number is positive, the answer is positive.

$$-10 + 13 = 3$$

Do one more:

$$16 + (-18) =$$

Again, the numbers have opposite signs, so we subtract the smaller number from the larger number ($18 - 16 = 2$). The sign of the larger number is negative, so the answer is negative.

$$16 + (-18) = -2$$

Looking Ahead

One important use of positive and negative numbers occurs when atoms form ions. Ions are atoms that have a positive or a negative charge. That charge is written in the upper right corner of the symbol for the element. For instance, you will learn that aluminum (symbol Al) forms a +3 ion which we write as Al^{+3} or Al^{3+}. Many textbooks write charges with the number before the sign (Al^{3+}), which we read as "aluminum three plus ion," rather than with the number after the sign (Al^{+3}). You will also learn that chlorine (symbol Cl) forms a -1 ion which will be written Cl^{-1} or Cl^{1-} or simply Cl^{-}. (Keep in mind that the number "one" is often omitted and merely "understood" to be there.)

Subtracting Signed Numbers

Subtracting signed numbers is a little more confusing for many people because we use the negative sign (–) in two ways: to show a number is negative and also as a subtraction or take away sign. Let's look at some examples in everyday life.

We encounter positive and negative numbers most often in our everyday life in our checking accounts. Every time you write a check, you take away a positive value from your account. Suppose you have 100 dollars deposited and you write a check for 10 dollars, you will have

$$100 - 10 = 90$$

Now suppose you don't write a check but instead the bank adds a 10 dollar service charge (–10) to your account. You now have:

$$100 + (-10) = 90$$

You can see that adding a negative 10 gives the same result as subtracting a positive 10. This same principle works in everyday life. For instance, there are two ways of disciplining children: taking away something they like (subtracting a positive) or giving them something they don't like (adding a negative).

The rule for subtraction can be summarized as: add the opposite of the second number to the first.

Let's try some examples.

$$6 - 12 =$$

We will put the positive sign on the 12 to explicitly show that the 12 is positive

$$6 - (+12) =$$

The opposite of 12 is –12, so we have

$$6 + (-12) = -6$$

Try another example.

$$-9 - 18 =$$

First rewrite the problem showing that the 18 is positive.

$$-9 - (+18) =$$

The opposite of 18 is –18, so we add

$$-9 + (-18) = -27$$

If you already owed nine dollars and someone took your credit card and gave you an eighteen dollar debt, it would have the same result as if that person took eighteen dollars from you.

Now suppose you owe sixteen dollars on your credit card. Then the bank tells you that it made a mistake and is taking away $9 of the debt. We can write this as

$$-16 - (-9) =$$

The opposite of –9 is 9, so we add

$$-16 + 9 = -7$$

You would then only owe $7.

Taking away a negative is comparable to adding positive. Another analogy from everyday life might help you understand: if a doctor removes a pain or a disease (a negative), the patient feels better (more positive).

There is another way to look at the subtraction of signed numbers. We can look at subtracting as finding how far apart two numbers are, which is called finding the difference of the numbers. Let's look at an example. If the temperature changed from 70° in the afternoon to 45° in the evening, how much did the temperature change from afternoon to evening? Finding the difference of two numbers always follows this pattern:

$$\text{Difference} = \text{To} - \text{From}$$

We can write this problem as a subtraction

$$45 - 70 =$$

$$45 + (-70) = -35$$

The temperature became 35 degrees colder in the evening.

I (Tim) once took a train trip across Canada in the middle of January from Winnipeg, where it was $-30°F$, to Vancouver, where it was $50°F$. What was the difference in temperature?

$$\text{Difference} = \text{To} - \text{From}$$

$$50 - (-30) = 50 + 30 = +80$$

It was 80 degrees warmer in Vancouver than it was in Winnipeg.

Looking Ahead

In the chapter on radioactivity you will study how radioactive atoms lose beta (β) particles. The nucleus of an atom has positive particles called protons and neutral particles with a zero charge called neutrons. When a radioactive nucleus becomes unstable, a neutron loses a negative particle called a beta particle. For instance when a carbon nucleus with a charge of 6 loses a beta particle with a -1 charge, its charge changes to 7.

$$+6 - (-1) = +7$$

$$6 + 1 = 7$$

Take away a negative 1 from 6 and it becomes 7. So when a nucleus loses a negatively charged beta particle, its charge always increases by one.

Multiplying and Dividing Signed Numbers

When you multiply two numbers or "take their product," the problem can be written several different ways:

$$2 \times 3 = 6 \qquad 2 \bullet 3 = 6 \qquad 2 * 3 = 6 \qquad 2(3) = 6$$

The form you will see most often is:

$$2(3) = 6 \qquad \text{or} \qquad (2)(3) = 6$$

where the numbers are simply written side-by-side separated by parentheses.

The product of two positive numbers is a positive number which you already know from grade school math.

$$2(10) = 20 \qquad 3(10) = 30 \qquad 4(10) = 40$$

The product of a positive number and a negative number is a negative number. If you double or triple, etc. a debt, you will have an even bigger debt.

$$2(-10) = -20 \qquad 3(-10) = -30 \qquad 4(-10) = -40$$

The rule that many people find a little strange at first is that the product of two negative numbers is a positive number. But if you think about it in ordinary language, the rule makes sense. If I take away two (–2) of your $10 debts (–10), you will be $20 richer (+20), if I take away 3 of your $10 debts, you will be $30 richer, etc.

$$-2(-10) = 20 \qquad -3(-10) = 30 \qquad -4(-10) = 40$$

Important Point

It's important you do not mix up the rules for multiplication of signed numbers with the rules for addition.

$$-3 + (-5) = -8 \qquad\qquad \text{negative + negative = negative}$$

but

$$-3(-5) = 15 \qquad\qquad \text{negative × negative = positive}$$

There is one multiplication you will see that may not look like a multiplication. If you see a negative sign in front of a set of parentheses, it means you multiply by negative one.

$$-(-5) \text{ means } -1(-5) = 5$$

$$-(-11) \text{ means } -1(-11) = 11$$

Note that multiplying by –1 simply changes the sign of the number.

The rules for division are exactly the same as for multiplication.

A positive number divided by a positive number gives a positive answer:

$$12 \div 2 = 6$$

A negative number divided by a positive number or a positive number divided by a negative number gives a negative answer:

$$(-12) \div 2 = -6 \qquad \text{or} \qquad 12 \div (-2) = -6$$

A negative number divided by a negative number yields a positive answer:

$$(-12) \div (-2) = 6$$

Picking up the Pieces
Fractions, Decimals and Percents

Fractions

We use fractions to show something is cut into equal pieces. The bottom of the fraction, called the denominator, tells us how many parts the whole is cut into. The top of the fraction, called the numerator, tells us how many of the parts we have. The line that separates them is called the fraction line or fraction bar.

$$\frac{\text{numerator}}{\text{denominator}}$$

The fraction $\frac{4}{5}$ tells us that something is cut into five equal pieces, and we have four of them. It might represent a pie cut into five equal pieces and there are four pieces left (Figure 2.1) or it could be your grade if you got four questions right on a five question quiz.

4/5 pie

Figure 2.1

A fraction can be expressed in words in several ways. We can say four out of five (as in 4 out of 5 doctors recommend...), four-fifths, four per five, 4 divided by 5 or as the ratio of 4 to 5.

Notice that $\frac{4}{5}$ of a pie is less than one whole pie. Any fraction smaller than one is called a proper fraction. Before a piece of the pie was eaten, we had $\frac{5}{5}$ or one whole pie. (Remember a fraction is another way of writing a division; therefore $\frac{5}{5}$ means $5 \div 5 = 1$.)

Important Point

Any number or variable or unit divided by itself equals one.

Examples: $\frac{5}{5} = 1$ $\qquad \frac{10}{10} = 1$ $\qquad \frac{x}{x} = 1$ $\qquad \frac{\text{inches}}{\text{inches}} = 1$

Fractions that are equal to one or larger, such as $\frac{9}{5}$, are called improper fractions.

You will see the word "per" often in chemistry problems. Any time you see "per," it means you have a fraction. If you drive 450 miles *per* 12 gallons of gasoline, you can write it as the fraction:

$$\frac{450 \text{ miles}}{12 \text{ gallons}}$$

The item following per is always in the denominator, the bottom of the fraction. This example also shows you the type of fraction you will see most often in chemistry, which will not be just plain numbers but will have "units" like gallons or miles.

Likewise if we drive 65 miles per hour, we can write the fraction:

$$\frac{65 \text{ miles}}{1 \text{ hour}}$$

Notice that if we have a fraction with units such as 65 miles per hour, we can write it as 65 miles/hour, $\frac{65 \text{ miles}}{\text{hour}}$ or $\frac{65 \text{ miles}}{1 \text{ hour}}$ where the third form emphasizes that a simple number such as 65 can also be written as a fraction with a denominator of 1. Then we have both a number and a unit in both numerator and denominator. We will talk more about fractions with units in Chapter 5.

Fractions are often written in textbooks with a slanted fraction line (/), mainly because it only takes one line of type. Thus 4/5 is the same as $\frac{4}{5}$. To make a fraction with a slanted line easier to work with, you may want to rewrite it with a horizontal fraction line.

Multiplication is the main operation you will perform with fractions in chemistry. Here are some examples of multiplying fractions:

$$\frac{\overset{1}{\cancel{2}}}{\underset{1}{3}} \times \frac{\overset{1}{3}}{\underset{8}{\cancel{16}}} = \frac{1}{8} \qquad \frac{\overset{1}{\cancel{4}}}{5} \times \frac{3}{\underset{1}{\cancel{4}}} = \frac{3}{5} \qquad \overset{20}{\cancel{2000}} \times \frac{1}{\underset{1}{\cancel{100}}} = 20$$

Notice how the numbers can be reduced before multiplying.

When you see a statement like "1/4 of" or "1/2 of," it means you are going to multiply.

$$\frac{1}{4} \text{ of } \frac{1}{2} \quad \text{means} \quad \frac{1}{4} \times \frac{1}{2} = \frac{1}{8}$$

If we cut 1/2 of a pie into 4 pieces, each piece will be 1/8 of the whole pie.

Likewise

$$\frac{1}{3} \text{ of } 9 \text{ means } \frac{1}{\cancel{3}} \times \frac{\cancel{9}^{3}}{1} = \frac{3}{1} = 3$$

One-third of nine dollars is three dollars. Notice that multiplying by the fraction 1/3 gives the same result as dividing by 3.

We can take any fraction or whole number and write its reciprocal. The reciprocal of a fraction is simply the fraction turned upside down.

The reciprocal of $\frac{9}{5}$ is $\frac{5}{9}$. The reciprocal of $\frac{7}{9}$ is $\frac{9}{7}$. The reciprocal of $\frac{1}{100}$ is $\frac{100}{1}$ or 100.

If we want to take the reciprocal of a whole number such as 100, we can always write the number with a denominator of 1, for instance $100 = \frac{100}{1}$. Then to find the reciprocal of 100, we just turn its fraction form upside down to get $\frac{1}{100}$.

If the fraction has units, we must move the units along with their numbers. The reciprocal of $\frac{5280 \text{ feet}}{1 \text{ mile}}$ is $\frac{1 \text{ mile}}{5280 \text{ feet}}$.

How many quarters are there in a $10 roll of quarters? We can write this as a division problem:

$$10 \div \frac{1}{4} = 40 \qquad \text{or} \qquad \frac{10}{\frac{1}{4}} = 40$$

which we can state in words as "How many $\frac{1}{4}$'s are there in 10?" You know from experience that there are 4 quarters in a dollar. So we can also write this problem as a multiplication

$$10 \times 4 = 40$$

You can see that multiplying by 4 (the reciprocal of $\frac{1}{4}$) gives the same result as dividing by $\frac{1}{4}$.

In general, to divide a number by a fraction, multiply the number by the reciprocal of the fraction. Let's look at a couple examples:

$$10 \div \frac{1}{10} = 10 \times \frac{10}{1} = 10 \times 10 = 100$$

$$\frac{3}{5} \div \frac{3}{10} = {}^{1}\frac{\cancel{3}}{\cancel{5}}_1 \times \frac{\cancel{10}^{2}}{\cancel{3}}_1 = 2$$

How many pennies are there in one dollar? Of course there are one hundred. We can write this as a division problem:

$$1 \div \frac{1}{100} = 1 \times \frac{100}{1} = 1 \times 100 = 100$$

How many quarter slices are there in one pie?

$$1 \div \frac{1}{4} = 1 \times \frac{4}{1} = 1 \times 4 = 4$$

Can you see the pattern? When you divide something into $\frac{1}{100}$'s, you have 100 pieces. When you cut something into $\frac{1}{4}$'s, you have 4 pieces. We will see this pattern again when we look at the metric system in Chapter 4.

Looking Ahead

In the metric system you will see this same pattern. One tenth of a meter is called a decimeter. If each decimeter is 1/10 of a meter, the meter has been cut into 10 equal pieces. So we can say:

1/10 meter = 1 decimeter   or   1 meter = 10 decimeters

Decimals

Often decimals are more convenient to use than fractions, especially when we use a calculator. Here is a decimal place value chart to refresh your memory.

thousands	hundreds	tens	ones	decimal point	tenths	hundredths	thousandths	ten-thousandths	hundred-thousandths	millionths	ten-millionths	hundred-millionths	billionths
3	7	3	5	.	2	5	7	8	2	6	4	2	9

You read the decimal 0.539 as five hundred thirty-nine thousandths which is equivalent to the fraction $\frac{539}{1000}$. The decimal 0.69 means sixty-nine hundredths and is equivalent to the fraction $\frac{69}{100}$. The decimal 0.000007 means seven millionths and is the same as the fraction $\frac{7}{1000000}$. You don't have to write the zero in front of the decimal point, but most of the time we put it there to help keep the decimal point from getting lost.

To change a fraction into a decimal, simply divide the denominator into the numerator.

$$\frac{3}{4} = 0.75 \qquad \frac{9}{5} = 1.8 \qquad \frac{4}{5} = 0.8 \qquad \frac{1}{3} = 0.333$$

In the first three examples above the division stops or terminates. In the fourth example the 3's repeat endlessly and we have to round the answer off. We will discuss rounding off in detail in the next chapter.

In the metric system you will often see fractions with multiples of ten in the denominator. A "short cut" for dividing these fractions is to move the decimal point of the numerator to the left as many places as the number of zeros in the denominator.

$$\frac{1}{10} = 0.1 \qquad \frac{1}{100} = 0.01 \qquad \frac{1}{1000} = 0.001 \qquad \frac{1}{1,000,000} = 0.000001$$

In the next chapter we will discuss multiplying and dividing decimals using a calculator. But you can multiply and divide decimals by multiples of ten in your head. To multiply a decimal by a multiple of 10, move the decimal point to the right a number of places equal to the number of zeros.

$$0.485 \times 1000 = 485 \qquad \text{move 3 places}$$

$$0.00057 \times 100 = 0.057 \qquad \text{move 2 places}$$

$$25 \times 10 = 250 \qquad \text{move 1 place}$$

To divide a decimal by a multiple of 10, move the decimal point to the left a number of places equal to the number of zeros.

$$\frac{0.485}{1000} = 0.000485 \qquad \text{move 3 places}$$

$$\frac{0.000057}{100} = 0.00000057 \qquad \text{move 2 places}$$

$$\frac{25}{10} = 2.5 \qquad \text{move 1 place}$$

Percents

Percents are special fractions that mean "out of one hundred." "Per" means "out of" and "cent" means one hundred, just as there are one hundred cents in a dollar and one hundred years in a century. Here are some examples demonstrating the relationships among percents, fractions, and decimals:

$$9\% = \frac{9}{100} = 0.09 \qquad 89\% = \frac{89}{100} = 0.89$$

Notice that, because we are dividing by 100 when we convert a percent to a decimal, we move the decimal point two places to the left.

If you have percents less than 1, they are handled the same way. Normal saline salt solution that is used in hypodermic injections is 0.9% salt. We can convert this percent to a decimal like this:

$$0.9\% = \frac{0.9}{100} = 0.009$$

In computing taxes, a *mill* is 0.1% of a dollar. We can convert this percent to a decimal:

$$0.1\% = \frac{0.1}{100} = 0.001$$

So a 1-mill levy will tax you 1 dollar out of every thousand dollars.

What is 100%? One hundred percent is one, the whole thing.

$$100\% = \frac{100}{100} = 1$$

So if you give something your whole effort, you are giving 100%.

We use percents when we want to have an easy way to see how big the part is compared to the whole. Thus if you answered 45 problems right out of a possible 60 on one test and 43 right out of 50 on the second test, it is not immediately obvious that you actually did better on the second test. On the first test you received

$$\frac{45}{60} \times 100\% = 75\%$$

while on the second test you received

$$\frac{43}{50} \times 100\% = 86\%$$

In general percents are calculated

$$\text{Percent} = \frac{\text{part}}{\text{whole}} \times 100\%$$

In glucose (a sugar), 72 grams out of 180 grams are carbon. What is the percent of carbon?

$$\text{Percent} = \frac{\text{part}}{\text{whole}} \times 100\% = \frac{72 \text{ gram}}{180 \text{ gram}} \times 100\% = 40\%$$

Suppose 180 students out of 200 passed the chemistry course. What percent passed?

$$\text{Percent} = \frac{\text{part}}{\text{whole}} \times 100\% = \frac{180 \text{ students}}{200 \text{ students}} \times 100 = 90\%$$

There is one other type of percent problem you may see. In this problem you are given a percent and a total and you want to find the part. Let's look at some examples.

If 20% of 2500 students are from out-of-state, how many students are from out-of-state? Remember 20% means the fraction $\frac{20}{100}$ and when we say "of," it means to multiply. So we write:

$$20\% \times 2500 = \frac{20}{100} \times 2500 = 500 \quad \text{or} \quad 0.20 \times 2500 = 500$$

In general we can write

$$\text{Part} = \frac{\%}{100} \times \text{Total} \quad \text{or} \quad \text{Part} = (\text{Percent as a decimal}) \times \text{Total}$$

What is 35% of 120?

$$\frac{35}{100} \times 120 = 42 \quad \text{or} \quad 0.35 \times 120 = 42$$

If you want to leave a 15% tip on a $55 meal, how much do you leave?

$$\text{Part} = \frac{15}{100} \times 55 = 8.25 \quad \text{or} \quad 0.15 \times 55 = 8.25$$

So you leave a little over $8.00.

The Little Big Numbers
Exponents

Whenever we multiply the same number by itself many times, we can shorten the writing required by using exponents. If we multiply 5 four times

$$5 \times 5 \times 5 \times 5$$

we can write it as

$$5^4$$

The 5 is the base; the 4 is the exponent. In words, we say 5 to the fourth power.

In chemistry numbers are often written in powers of 10 with a base of "10" like this:

$$10^3$$

We read 10^3 as 10 to the third power or 10 cubed,

$$10^3 = 10 \times 10 \times 10 = 1000$$

Now look at the following pattern:

$$10^1 = 10$$
$$10^2 = 10 \times 10 = 100$$
$$10^3 = 10 \times 10 \times 10 = 1000$$
$$10^4 = 10 \times 10 \times 10 \times 10 = 10,000$$
$$10^5 = 10 \times 10 \times 10 \times 10 \times 10 = 100,000$$
$$10^6 = 10 \times 10 \times 10 \times 10 \times 10 \times 10 = 1,000,000$$

You can see from the pattern that the exponent is equal to the number of zeros. Once you get used to exponents, it is much easier to write 10^{12} than it is to write 1,000,000,000,000. It will be helpful to you to become familiar with the names of the exponential forms of key powers of ten because they are used so often.

$$10^3 = 1 \text{ thousand} = 1000$$

$$10^6 = 1 \text{ million} = 1,000,000$$

$$10^9 = 1 \text{ billion} = 1,000,000,000$$

$$10^{12} = 1 \text{ trillion} = 1,000,000,000,000$$

The Hourglass of Our Lives

For most of us, large numbers like those just mentioned do not have much meaning. When we hear news reports about billion dollar government budgets, most of us go numb. To give you an idea of the relative size of these numbers, here is how long each of these numbers is in seconds:

10^3 seconds = 16 minutes and 40 seconds

10^6 seconds = 11 days and 13 hours and 47 minutes

10^9 seconds = 31 years and 35 weeks and 5 days

10^{12} seconds = 31 thousand years (back to the cave person days)

If you know someone turning 31 years old, wish them a happy 1 billion seconds!

Negative Exponents

Exponents can also be negative. A negative exponent means to take the reciprocal of the base and give the base a positive exponent. The negative exponents of 10 follow this pattern:

$$10^{-1} = \frac{1}{10} = 0.1$$

$$10^{-2} = \frac{1}{10^2} = \frac{1}{100} = 0.01$$

$$10^{-3} = \frac{1}{10^3} = \frac{1}{1000} = 0.001$$

$$10^{-4} = \frac{1}{10^4} = \frac{1}{100,000} = 0.0001$$

$$10^{-5} = \frac{1}{10^5} = \frac{1}{100,000} = 0.00001$$

$$10^{-6} = \frac{1}{10^6} = \frac{1}{1,000,000} = 0.000001$$

$$10^{-9} = \frac{1}{10^9} = \frac{1}{1,000,000,000} = 0.000000001$$

For negative exponents, the exponent is equal to the number of decimal places. Negative exponents are convenient for small numbers. Compare 10^{-12} to 0.000000000001! Keep in mind the larger the negative exponent, the smaller the number actually is.

One very important exponent is zero. $10^0 = 1$. In mathematics anything to the zero power is 1. Thus we can say $2^0 = 1$; $x^0 = 1$; etc.

To get an idea why, let's continue the pattern for zeros on the last page:

$10^4 = 10,000$ 4 zeros

$10^3 = 1000$ 3 zeros

$10^2 = 100$ 2 zeros

$10^1 = 10$ 1 zero

$10^0 = 1$ no zeros

We will show you another reason why $10^0 = 1$ shortly.

Multiplying and Dividing Exponential Numbers

When you multiply two exponential numbers with the same base, you simply add the exponents. Let's look at how this works:

$$10^3 \times 10^4 = (10 \times 10 \times 10) \times (10 \times 10 \times 10 \times 10) = 10^{3+4} = 10^7$$

As you can see, a total of seven 10's are being multiplied. Adding the exponents is the shortcut.

Note: the multiplication sign between the parentheses is often omitted and the problem written

$$(10 \times 10 \times 10)(10 \times 10 \times 10 \times 10) =$$

This rule works whether the exponents are positive or negative. We'll do some examples using the rules we learned in the last section for adding signed numbers.

$$10^{-3} \times 10^{-9} = 10^{-3+(-9)} = 10^{-12}$$

$$10^{-4} \times 10^8 = 10^{-4+8} = 10^4$$

$$10^{-8} \times 10^{-8} = 10^{-8+(-8)} = 10^{-16}$$

$$10^{-9} \times 10^{23} = 10^{-9+23} = 10^{14}$$

Now let's see if we can prove to ourselves that 10^0 really is equal to 1.

$$10^{-1} \times 10^1 = 10^{-1+1} = 10^0$$

or, if we convert these exponential numbers to fractions we get

$$\frac{1}{\cancel{10}_1} \times \frac{\cancel{10}^{\;1}}{1} = 1$$

So 10^0 must equal 1.

When we divide exponential numbers, we subtract the exponent of the denominator from the exponent of the numerator. Let's see how this works. (Remember that any number divided by itself is equal to one. $10/10 = 1$)

$$\frac{10^4}{10^7} = \frac{\cancel{10} \times \cancel{10} \times \cancel{10} \times \cancel{10}}{\cancel{10} \times \cancel{10} \times \cancel{10} \times \cancel{10} \times 10 \times 10 \times 10} = \frac{1}{10 \times 10 \times 10} = \frac{1}{10^3} = 10^{-3}$$

The four 10's in the numerator cancel four of the seven 10's in the denominator. We get the same result by subtracting the exponents

$$10^{4-7} = 10^{4+(-7)} = 10^{-3}$$

Let's try some more:

$$\frac{10^{-3}}{10^{-4}} = 10^{-3-(-4)} = 10^{-3+4} = 10^1 = 10$$

$$\frac{10^{-3}}{10^4} = 10^{-3-4} = 10^{-3+(-4)} = 10^{-7}$$

$$\frac{10^6}{10^9} = 10^{6-9} = 10^{6+(-9)} = 10^{-3}$$

If You Get Tired of Counting Zeros
Scientific Notation

You will use scientific notation often in your chemistry class. A number written in scientific notation is simply an ordinary decimal number between 1 and 10, called the coefficient, multiplied by a power of ten.
Here is an example:

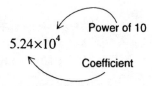

The decimal form of 5.24×10^4 is 52,400 because we can write 10^4 as 10,000 and multiply: $5.24 \times 10^4 = 5.24 \times 10,000 = 52,400$.

The exponent can also be negative:

$$2.39 \times 10^{-4}$$

The decimal form of 2.39×10^{-4} is 0.000239 because we can write 10^{-4} as 0.0001 and multiply:

$$2.39 \times 10^{-4} = 2.39 \times 0.0001 = 0.000239$$

In the next chapter you will learn to multiply and divide using scientific notation as well as convert between scientific notation and decimal form quickly and easily using your calculator.

But now let's convert some numbers from decimal form to scientific notation without a calculator just so we can get a feel for the numbers. To convert a decimal number to scientific notation, move the decimal point until you get a number between 1 and 10. The exponent will be the number of places you moved the decimal point. The sign of the exponent will depend on whether it is a small number (negative exponent) or a big number (positive exponent).

$$120,000 = 1.2 \times 10^{5}$$

5 places   + exponent: large number

If you don't see a decimal point, assume it comes at the end of the number.

$$0.000027 = 2.7 \times 10^{-5}$$

5 places   – exponent: small number

$$1,870,000,000 = 1.87 \times 10^{9}$$

9 places   + exponent: large number

$$0.0000000905 = 9.05 \times 10^{-8}$$

8 places   – exponent: small number

Scientific notation puts unwieldy large or small numbers into a form that is much easier to use. Sometimes you will see regular size numbers written in scientific notation.

$$18 = 1.8 \times 10^{1}$$

$$0.19 = 1.9 \times 10^{-1}$$

As you can see, numbers like these are easier to write in decimal form.

To convert a number written in scientific notation to decimal form, move the decimal point of the coefficient a number of places equal to the exponent, adding zeros as necessary. If the exponent is positive, move the decimal point to the right to make a large number. If the exponent is negative, move the point to the left to make a small number. Let's look at some examples:

$$2.37 \times 10^5 = 237,000 \qquad \text{move 5 places to the right}$$

$$6.59 \times 10^{-4} = 0.000659 \qquad \text{move 4 places to the left}$$

Looking Ahead

In the chapter on moles, later in the class, you will work with a number called Avogadro's number. You will see this number so much you'll probably get sick of it. Avogadro's number is 6.02×10^{23}. It is the number of atoms in a mole. (Just like 12 is the number of eggs in a dozen.) If we write it in the decimal form, it looks like this:

$$602,000,000,000,000,000,000,000$$

You can see that it is much easier to write it in scientific notation. Avogadro's number is unbelievably big. It is 602,000 billion billion. That is 602 thousand times a billion times a billion. Avogadro's number of seconds is190 trillion centuries!

Now try the practice problems on page 167.

Good Study Habits

Good study habits are not among our "basic instincts." Most of us have to work long and hard to develop good study habits. Here are some concrete things you can do to promote your success as a student.

- Set a long-term goal, that is, a career goal. Goals keep us going when the going gets rough.

- Set short-term goals, that is, daily or weekly goals. Short term goals motivate us and keep us on target.

- Make a realistic daily, weekly, and term-long schedule. Post copies of your daily schedule where others can see it. That will force you to stick to it. Also schedule relaxation time for yourself! We all need that!

- Observe "good students" and make yourself look and act like a good student. Soon you will start thinking of yourself as one. We often have to "act" ourselves into a different way of thinking.

🍂   Create a study area that meets your needs. The study area should have few distractions and be quiet. Try to keep it uncluttered. Having "a place" to study helps you get in the mood to study.

🍂   Select the best time for you to study. We all have different "internal clocks." Study when you are most alert.

🍂   Reflect and review daily. Reading a lesson and doing problems is not studying. You read the lesson and do the problems to gain the knowledge you need to study the material. Studying involves moving material from memory to understanding. If you can't explain a concept in your own words and relate it to something you already know, you probably don't understand it and won't remember it very long.

🍂   Get to know others in your class and form a serious study group. Then study together at least once a week. It will keep you "honest." It is easy to slip into sloppy study habits and fool yourself into thinking you know more than you do. Having to explain ideas in a study group will bring you back to reality very quickly. We learn and learn how to learn from working with others.

🍂   Work to stay relaxed while you are studying. For some pointers, see Learning to Relax at the end of Chapter 7.

CHAPTER 3

Push Button Thinking Using Your Calculator

Probably the best friend you have in a chemistry class is your calculator. If you use your calculator correctly, it will save you precious seconds during a test. It will make doing homework problems outside class much more efficient and enjoyable, and allow you to spend more time on the concepts rather than laboring over the calculations. It is much more important to understand the concepts of chemistry than to do the calculations.

Ten Dollar Calculators That Work

Many students have difficulty in the first weeks of the semester because they don't have the type of calculator they need to quickly and efficiently perform the tasks the instructor and textbook will require of them. You may sit down in your first class with a four-function calculator that just does the basic operations of adding, subtracting, multiplying and dividing—the type they give you when you open a checking account. This type of calculator is not powerful enough to do many of your chemistry problems. On the other hand, you don't want to purchase an engineering calculator with so many functions it is intimidating to look at—and has a manual as thick as *War and Peace*!

You need a scientific calculator that will help you quickly and easily dispatch the problems set before you in your class, especially on tests. The calculator should have the following capabilities:

- The four basic arithmetic functions.

- A button to enter scientific notation (EE or EXP on most calculators).

- Buttons to easily enter logarithms. (Don't worry about what logarithms are right now. We will discuss these later in Chapter 16.)

- Buttons to convert between scientific notation and decimal form.

- Buttons to do statistical calculations. (You won't need them in this class, but many of you will probably take statistics later in your program. You may also need to do some simple statistics in some of your more advanced nursing classes.)

- At least 10 places on the display so you can display fairly large numbers in decimal form. (Called 10 + 2 on some models.)

Below is a list of calculators that are inexpensive (under $10) and do the job you need a calculator to do.

- Casio fx-250HA or fx-250HB

- Texas Instruments TI-30X or TI-30X Solar

The First Five Buttons
The Basic Operations

Nothing is easier than multiplying or dividing on a calculator. If you want to multiply 2.71×1.91, you simply punch

2.71 ⨯ 1.91 = The display reads *5.1761*

When you divide two numbers, the process is the same. Let's divide 3.75 by 1.2. This problem can be presented to you as $3.75 \div 1.2$ or $3.75/1.2$ or $\frac{3.75}{1.2}$. No matter how the division is written, it is entered into the calculator as

3.75 ÷ 1.2 = The display reads *3.125*

Most of the time in chemistry you will do several operations in one problem. An example of this type of problem is

$$2.875 \times \frac{1}{3.125} =$$

Many students enter the operations like this:

2.875 ⨯ 1 ÷ 3.125 = The display reads *0.92*

It is much simpler and less prone to error if the multiplication step is left out when all you are doing is multiplying by one. You simply enter

2.875 (+) 3.125 (=)   The display reads   *0.92*

Later, when you do conversions, you will do many multiplications and divisions in one problem. Let's try one:

$$\frac{2.375 \times 4.15}{0.03125 \times 1.577} =$$

Many students do the problem like this:

First they multiply the numbers in the numerator

2.375 (×) 4.15 (=)   The display reads   *9.85625*

Then they write this answer down, push (C) to clear the display and then multiply the numbers in the denominator:

0.03125 (×) 1.577 (=)   The display reads   *0.04928125*

Then they write this number down, clear the display, reenter 9.85625, and finally divide by 0.04928125 to get the answer 200.

In other words, they multiply the numbers in the numerator first, erase the answer, then multiply the numbers in the denominator, reenter the first answer and then do the division.

There is an easier way to do it. Do the multiplications and divisions in order from left to right. Press (×) before numbers that are in a numerator (after the first number) and press (÷) before any number in a denominator.

2.375 (×) 4.15 (÷) 0.03125 (÷) 1.577 (=)   The display reads   *200.*

You can change the order, as long as you press (×) before a number in the numerator and press (÷) before a number in the denominator.

2.375 (÷) 0.03125 (÷) 1.577 (×) 4.15 (=)   The display reads   *200.*

Let's try another:

$$14.4 \times \frac{1}{1.6} \times \frac{13.3}{20} =$$

We punch:

14.4 (÷) 1.6 (×) 13.3 (÷) 20 (=)   The display reads   *5.985*

Doing problems this way can save you much time on test questions.

What Do I Do With All These Numbers?
Rounding Off and Significant Figures

If you divide 2 by 7 on your calculator, the display reads 0.285714286. Now ten decimal places can be quite cumbersome. Fortunately, as you know, you do not have to write all those numbers when you answer a problem. You can round off your answer. Answers beyond a certain number of places are not just cumbersome, they are misleading. So we round our answers off to a reasonable number of decimal places.

In order to round off, we have to understand significant figures (sometimes referred to as sig figs). In chemistry (and in all science) we do not work with pure numbers like in algebra class. We work with measurements. If you take your body temperature with a fever thermometer, it may look like Figure 3.1. The mercury stops between 98 and 99 degrees. It looks like it is between 98.2 and 98.4 degrees (each small line on the thermometer is 0.2 of one degree), so we might guess 98.3. In all measurements the last digit is a good guess and your guess might be a little different from mine. This temperature (98.3 degrees) has three significant figures. If we round it off to two places, it will have two significant figures (98 degrees).

Figure 3.1

Some laboratory balances read to the hundredths place. The digital readout of an object's mass might look like this: *14.49*. This reading has four significant figures. Other (more expensive) balances read to the ten-thousandths place. The digital display on this balance might read *14.4927* for this object. This reading has six significant figures.

What Are the Rules for Significant Figures?

How can you tell if digits are significant? Most of the confusion involves the number zero. Your textbook will give you 4 or 5 rules something like these:

1.   All non-zero digits are significant.

This means that any number that is not zero is significant. 7.121486 has 7 significant figures; 4215 has 4 significant figures.

2.   All zeros between non-zero digits are significant.

7002 has 4 significant figures; 7.1002 has 5 significant figures; 702 has 3 significant figures; 80.003 has 5 significant figures.

3.   All zeros to the left of the first non-zero digit are not significant.

Here is where your normal everyday definition of "significant" may clash with the chemistry definition. The number 0.0025 has only 2 significant figures, as the zeros to the left of the 2 are not significant. That does not mean they don't have to be there. If you took them away, you would get .25, a very different number! A common dosage of Vasotec, a blood pressure-lowering drug, is 0.0025 grams. If you ignored the zeros, the dosage would become .25 grams, 100 times too high and probably fatal to the patient. Those zeros then have the very important job of being place holders; however, they are not significant in the sense of expressing the accuracy of the number.

If we convert 0.0025 kilograms to grams, it becomes 2.5 grams. Because we changed the units, we no longer needed the left-hand zeros as place holders.

4.   When a decimal point is present, zeros to the right of the last non-zero digit are significant.

Thus 0.2300 has 4 significant figures, 2.4900 has 5 significant figures and 2.0 has 2 significant figures.

5.   If there is no decimal point written, you cannot tell if the zeros to the right are significant. This rule can confuse us if we are not careful.

For instance, we cannot say how many significant figures whole numbers like 8000 and 400 have. They have one significant figure; beyond that we can not say.

Now let's look at some examples that involve all the rules.

7.0010 has 5 significant figures. (All the zeros are significant because two are between non-zero digits and the last is in a number that has the decimal point written.)

0.0005 has 1 significant figure. (The zeros to the left are not significant.)

0.000500 has 3 significant figures. (The zeros to the left of the 5 are place holders but are not significant. The zeros to the right are significant because a decimal point is shown.)

1.000500 has 7 significant figures. (All the zeros are significant because the three zeros are between non-zero digits and the last two are to the right in a number that has the decimal point written.)

7000 has 1 significant figure. (No decimal point is shown so you cannot tell whether or not the zeros are significant.)

7000. has 4 significant figures. (This time the decimal point is shown.)

Scientific Notation to the Rescue

Let's look at the number 800. Suppose it represents 800 pounds. If 800 pounds is just a good guess at the weight of an object, then the number has only 1 significant figure. However, if you weighed the object and it weighed exactly 800 pounds, the number would have 3 significant figures. If you want to show that the number has 3 significant figures, you would have to write it with a decimal point included, that is, 800. lbs. But it is easy for a decimal point at the end of a whole number to "get lost." However, we can easily show the number of significant figures in numbers such as this by converting them to scientific notation. If you convert 800 to scientific notation, it becomes

$$8 \times 10^2$$

If you want to show that it has 3 significant figures, you can write it as

$$8.00 \times 10^2$$

When you write a number in scientific notation, the number of significant digits is expressed by the coefficient.

7000 lbs becomes	7×10^3	with 1 significant figure
7000 lbs becomes	7.0×10^3	with 2 significant figures
7000 lbs becomes	7.00×10^3	with 3 significant figures

$$7000 \text{ lbs becomes} \qquad 7.000 \times 10^3 \text{ with 4 significant figures}$$

Remember that the exponent plays no part in determining the number of significant figures. It merely tell us where to put the decimal point.

Rules for Rounding Off

You may remember these rules from a math class in high school. To round off a number to a certain place, you look at the number in the next place to its right. If the number to its right is less than 5, that is, if it is 0, 1, 2, 3, or 4, you leave the number alone. This is called rounding down. If the number to its right is 5 or greater, that is, if it is 5, 6, 7, 8, or 9, you increase the value of the number by one. This is called rounding up.

Let's round some numbers.

7.134 rounded to the hundredths place becomes 7.13 since the number in the thousandths place is 4, a number less than 5.

0.21029 rounded to the ten-thousandths place becomes 0.2103 since the number to the right of that place is 9, a number that is 5 or greater.

Sometimes when you round a number off it makes it a little easier to visualize the size of the number. The concentration of salt in normal saline solution, used in hypodermic solutions, is 0.9%. If we round the percent off to the one's place, it becomes 1% or about 1 part of salt in one hundred parts of solution.

Looking Ahead

Periodic tables in chemistry texts report the atomic masses of different elements with values from 4 to 7 significant figures. Sometimes in problems you will do later, it's helpful to round the atomic masses to the tenths place. For example:

Carbon (atomic mass 12.011) rounded to the tenths place becomes 12.0.

Hydrogen (atomic mass of 1.008) rounded to the tenths place becomes 1.0.

Aluminum (atomic mass 26.98) rounded to the tenths place becomes 27.0.

Sulfur (atomic mass of 32.06) rounded to the tenths place becomes 32.1.

Significant Figures in Multiplication and Division

When you multiply or divide measurements, your answer is limited by the number with the least number of significant figures. Suppose three students measure a box. Jane measures the length to be 3.12 cm, Carl measures the width to be 3.1 cm, and Donna measures the height to be 7.62 cm. To get the volume we multiply the three sides together.

$$3.12\,cm \times 3.1\,cm \times 7.62\,cm \ = \ 73.70064\,cm^3$$

However, we must round the calculated answer to 74 cm^3 because Carl made his measurement only to 2 significant figures.

When we calculate density, we divide mass by volume. That is

$$Density \ = \ \frac{mass}{volume}$$

Let's say we measure a sample's mass to be 7.4219 grams and its volume to be 1.23 mL. We divide 7.4219 by 1.23

$$\frac{7.4219\ g}{1.23\ mL}$$

the calculator answer is 6.034065041

Because 1.23 mL has only 3 significant figures, we must round the answer to 6.03.

Significant figures are important, but don't get too bogged down with them. They will come naturally to you as you do more calculations.

When Do I Hit the EXP Button? Scientific Notation on the Calculator

Calculators make scientific notation easy. You just need to have the right calculator and know which buttons to push.

Important
Point

Scientific calculators have a button that is marked (EE) (on most Texas Instruments calculators) or (EXP) (on most other brands). This is the button that lets you enter a number in scientific notation.

Let's take the number $\qquad$ 7.45×10^8

To enter this number, we enter the coefficient 7.45, then press (EXP). Two zeros appear at the right (smaller in size). These zeros are holding the places for the two digits of the exponent. (You will never need more than two.) We then enter the exponent 8. The display should look like this:

$$7.45 \quad 08$$

When you see this display, you read it as 7.45×10^8. Press (C) to clear the display and let's enter it again.

7.45 (EXP) 8.   The display reads   *7.45   08*

Enter 9.992×10^{12}

Press 9.992 (EXP) 12   The display reads   *9.992 12*

A common mistake students make is to hit (×) before (EXP). Don't do this! The "times 10 to the" is included in (EXP).

So you do **not** enter

7.45 (×) (EXP) 8 or 7.45 (×) 10 (EXP) 8

you simply press 7.45 (EXP) 8

Let's enter 9.12×10^{12}

Press 9.12 (EXP) 12   The display reads   *9.12 12*

Let's enter a number with a negative exponent: 7.216×10^{-9}

Enter 7.216 (EXP) 9 (+/−) or 7.216 (EXP) (+/−) 9   The display reads   *7.216 -09*

(+/−) toggles a number between negative and positive. If you press (+/−) after (EE) or (EXP), it toggles the exponent between negative and positive. Press (+/−) a couple of times and see how the sign of the exponent changes back and forth.

(+/−) is **not** the same as (−). (+/−) toggles between negative and positive. (−) is used only for subtraction.

Let's enter a few more numbers in scientific notation.

Enter the mass of an electron which is 1.6606×10^{-24} kilograms.

Press 1.6606 (EXP) 24 (+/−)   The display reads   *1.6606 -24*

Enter 10^5

When entering an exponential number where no coefficient is shown, you must supply the missing coefficient of 1. We can write 10^5 as 1×10^5.

Press 1 (EXP) 5   The display reads   *1. 05*

Enter 10^{-7}

Press 1 (EXP) 7 (+/-)   The display reads $1. \ ^{-07}$

Multiplication and Division

If you multiply two numbers in scientific notation, just press (×) between the numbers. Let's multiply

$$(2.12\times10^{4})(7.89\times10^{-9}) =$$

Enter 2.12 (EXP) 4 (×) 7.89 (EXP) 9 (+/-) (=)

The display reads $1.67268 \ ^{-04}$
Since each number has three significant figures, we round the answer to 1.67×10^{-4}. Note the only time you enter (×) is between the two numbers.

Let's divide two numbers:

$$\frac{8.81\times10^{4}}{7.472\times10^{8}} =$$

8.81 (EXP) 4 (÷) 7.472 (EXP) 8 (=)   The display reads $1.1790685 \ ^{-04}$

Since the least number of significant figures is 3, we round our answer to 1.18×10^{-4}.
Let's try another:

$$\frac{10^{-14}}{10^{-9}} =$$

Remember to supply the missing coefficients of 1.

Press 1(EXP) 14 (+/-) (÷) 1 (EXP) 9 (+/-) (=)

The display reads $1.0000000 \ ^{-05}$ which is written 1×10^{-5} or just 10^{-5}.

Now let's look at one last example that uses Avogadro's number (6.02×10^{23}), a number that you will see later in the class.

$$2.50\times10^{15} \times \frac{12.0}{6.02\times10^{23}} =$$

Press 2.5 (EXP) 15 (×) 12 (÷) 6.02 (EXP) 23 (=)

The display reads $4.9833887\ ^{-08}$. Since each of the three numbers has three significant figures, we round the answer to 4.98×10^{-8}.

Switching Between Scientific Notation and Decimal

Sometimes you may want to convert an answer back and forth between scientific notation and decimal form. This will be most important to you on exams, especially multiple choice exams where the answer choices could be written in either form.

The answers to problems in the back of your textbook can also be given in either scientific notation or decimal form. Often students are confused when they look up a hard fought answer in the answer key and discover their answer looks wrong. Their answer may not be wrong. The textbook may have given the answer in scientific notation while the student calculated it in decimal form.

Texas Instruments Calculators

On Texas Instrument calculators you use (FLO) and (SCI) along with (2nd) to convert between scientific notation and decimal form. On TI-30X calculator, (FLO) and (SCI) are second functions written in blue above (4) and (5). You use (2nd) to access these operations.

Pushing (2nd) (FLO) puts the calculator into normal decimal mode. (FLO comes from "floating point," computer programmers' jargon for a decimal number). In this mode, any number you enter will change to decimal form as soon as you hit another key (as long as the number will fit the display).

Pushing (2nd) (SCI) puts the calculator into scientific notation mode. Any number you enter will change to scientific notation as soon as you hit another key.

First let's change some numbers in scientific notation to decimal form.

Suppose you got 7.7×10^{-4} as an answer and want convert it to decimal form.

Enter 7.7 (EE) 4 (+/-)	The display reads	$7.7\ ^{-04}$
Press (2nd) (FLO)	The display reads	0.00077

Now let's convert 8.449×10^{7} to decimal form.

Enter 8.449 (EE) 7	The display reads	$8.449\ ^{07}$

Press [2nd] [FLO]   The display reads   *84490000*

Now let's try to convert 6.419×10^{-17} to decimal form.

Enter   6.419 [EE] 17 [+/-]   The display reads   *6.419* *-17*

Now press [2nd] [FLO]   The display still reads   *6.419* *-17*

Nothing happens because a number in scientific notation with a −17 exponent has too many decimal places to fit into the 10 spaces on the calculator. A number in scientific notation must have an exponent of 9 or less, either plus or minus, to fit into the ten decimal places on a calculator. That is, a number must be between nine billion and nine-billionth.

Now let's take some decimal numbers and convert to scientific notation.

Suppose you got 0.000788 for an answer and want to change it to scientific notation.

Enter 0.000788   The display reads   *0.000788*

Press [2nd] [SCI]   The display reads   *7.88* *-04*

Convert 6,992,000 to scientific notation.

Enter 6,992,000   The display reads   *6992000*

Press [2nd] [SCI]   The display reads   *6.992* *06*

You can press [2nd] [FLO] to convert back to normal decimal form.

Important Point

If you press [2nd] [SCI] to convert an answer to scientific notation, the calculator remains in that mode until you press [2nd] [FLO] to put the calculator into normal decimal mode. If you enter a decimal number, it will change to scientific notation as soon as you press another function key such as [×], [÷] or [=].

Likewise if you are in decimal mode when you enter a number in scientific notation, it will change to decimal form as soon as you hit another key (as long as the decimal form will fit in the display). In either of these cases the number is not affected, only the form of the number.

Let's look a few examples.

Press $\boxed{\text{2nd}}$ $\boxed{\text{SCI}}$ to put the calculator in scientific notation mode

Enter 0.0059   The display reads   *0.0059*

Press $\boxed{\times}$ or $\boxed{\div}$ The display changes to *5.9000000 ⁻03*

Press $\boxed{\text{C}}$ to clear the display. Now press $\boxed{\text{2nd}}$ $\boxed{\text{FLO}}$ to change your calculator to normal decimal mode.

Now enter 0.0059   The display reads   *0.0059*

Now press $\boxed{\times}$ or $\boxed{\div}$ The display still reads   *0.0059*

Casio, Sharp and Most of the Rest

Most brands of calculators have a $\boxed{\text{MODE}}$ button, usually in the upper left hand corner next to $\boxed{\text{INV}}$. They also have a table printed under the readout display that looks something like this:

MODE	•	0	4	5	6	7	8	9
	SD	COMP	DEG	RAD	GRA	FIX	SCI	NORM

or this

MODE	0:COMP	4:DEG	5:RAD	6:GRA
	• SD	7:FIX	8:SCI	9:NORM

The two numbers you need in this table to convert between scientific notation and decimals are 8:SCI (scientific notation) and 9:NORM (for normal decimal form).

Pressing $\boxed{\text{MODE}}$ $\boxed{9}$ puts the calculator into normal decimal mode. In this mode, a number you enter will change to decimal form as soon as you hit another key (as long as the number will fit in the display).

Pressing $\boxed{\text{MODE}}$ $\boxed{8}$ $\boxed{8}$ puts the calculator into scientific notation mode. Any number you enter will change to scientific notation as soon as you hit another key. The second number you enter after pressing $\boxed{\text{MODE}}$ $\boxed{8}$ tells the calculator how many places of the coefficient the calculator will display. For instance, if you press $\boxed{\text{MODE}}$ $\boxed{8}$ $\boxed{4}$, the answer will be rounded to 4 digits. It's better to let the calculator display 8 digits, then you can choose the proper number of significant figures for an answer.

Let's change some numbers from scientific notation to normal decimal form.

Suppose you got 7.7×10^{-4} as an answer and want to change it to decimal form.

Enter   7.7 (EXP) 4 (+/-)   The display reads $7.7^{\ -04}$

Press (MODE) (9)   The display reads 0.00077

Convert 2.568×10^{8} to decimal form.

Enter 2.568 (EXP) 8   The display reads $2.568^{\ 08}$

Press (MODE) (9)   The display reads 256800000

Now let's try to convert 7.489×10^{-13} to decimal form.

Enter 7.489 (EXP) 13 (+/-)   The display reads $7.489^{\ -13}$

Now press (MODE) (9)   The display still reads $7.489^{\ -13}$

Nothing happens because a scientific notation number with a −13 exponent has too many decimal places to fit into the 10 spaces on the calculator. A number in scientific notation must have an exponent of 9 or less, either plus or minus, to fit into the ten decimal places on a calculator. That is, a number must be between nine billion and nine-billionth.

Now let's convert decimal numbers to scientific notation.

Suppose you got 0.00489 as an answer and want to change it to scientific notation.

Enter 0.00489   The display reads 0.00489

Press (MODE) (8) (8)   The display reads $4.8900000^{\ -03}$

Convert 2,090,000 to scientific notation.

Enter 2,090,000   The display reads 2090000

Press (MODE) (8) (8)   The display reads $2.090000^{\ 06}$

Press (MODE) (8) (8) to put the calculator in scientific notation mode.

Important
Point

If you press [MODE] [8] [8] to convert an answer to scientific notation, the calculator remains in that mode until you press [MODE] [9] to change to normal decimal mode. If you enter a decimal number, it will change to scientific notation as soon as you press another function key such as [×], [÷] or [=].

Likewise, if you are in decimal mode when you enter a number in scientific notation, it will change to decimal form as soon as you press another key (as long as the decimal form will fit in the display). In either of these cases the number is not affected, only the form of the number.

Let's look at some examples.

Press [MODE] [8] [8], this puts the calculator in scientific notation mode:

Enter 0.0059   The display reads   *0.0059*

Press [×] or [÷]   The display changes to *5.9000000* ⁻⁰³

Press [C] to clear the display. Now press [MODE] [9] to return your calculator to normal decimal mode.

Now enter 0.0059   The display reads   *0.0059*

Now press [×] or [÷]   The display still reads   *0.0059*

Does the Answer Make Sense?

A calculator is a very handy instrument. It will make computations quick and easy. But it is only as good as the numbers you enter. Computer programmers have an acronym for this: GIGO. Garbage in, garbage out. If you are in a hurry on a test, it is easy to push the wrong button and not notice it. Let's look at some calculations and see if we can tell if the answer makes sense.

$$\frac{9.2\times10^8}{5.5\times10^5} = 1.7\times10^3$$

We can check our answer to see if it's in the right ballpark by looking at our exponents: $10^8 \div 10^5 = 10^3$. Also notice that 9.2 is a little less than twice 5.5, so the ratio of 1.7 is reasonable. Notice that the numerator is larger than the denominator so the answer has a positive exponent.

Let's look at another example:

$$\frac{7.9 \times 10^{-4}}{6.9 \times 10^{-9}} = 1.1 \times 10^5$$

We look at the exponents: $10^{-4} \div 10^{-9} = 10^{-4-(-9)} = 10^{-4+9} = 10^5$.

It checks out. In this case we divided a larger number by a smaller number and we get a number greater than one. Also notice that 7.9 is just a little larger than 6.9, so a ratio of 1.1 makes sense.

If by mistake we had hit ⊠ instead of ⊞, we would have gotten the wrong answer 5.5×10^{-12}. By quickly checking the exponents we can spot this kind of error. The few seconds you spend checking your answers is time well spent.

Remember that studies show that most students can raise their grades one or even two letters by doing nothing more than being neat and checking their work.

Important Point

Look at the following problem and the **incorrect** answer given.

$$\frac{7.50 \times 10^9}{6.5 \times 10^5 \times 3.5 \times 10^3} = 4.0 \times 10^7$$

Do a quick check of the exponents. Did your check tell you that the exponent in the answer should be around 10^1 and not 10^7? ($10^9 \div 10^5 \div 10^3 = 10^{(9-5-3)} = 10^1$)

Can you figure out what the student did wrong to arrive at the **incorrect** answer?

Doesn't it appear that the student carried out the calculation **incorrectly** as:

7.5 ⌈EXP⌉ 9 ⌈÷⌉ 6.5 ⌈EXP⌉ 5 ⌈×⌉ 3.5 ⌈EXP⌉ 3 ⌈=⌉

Instead of the correct key sequence:

7.5 ⌈EXP⌉ 9 ⌈÷⌉ 6.5 ⌈EXP⌉ 5 ⌈÷⌉ 3.5 ⌈EXP⌉ 3 ⌈=⌉

Remember that on page 30, we noted that all numbers in the denominator must be preceded by ⌈÷⌉ even if a × sign appears in front of the number. This student got the wrong answer because she forgot this rule.

Important Point

On multiple choice tests, the wrong answer you're likely to get is often one of the choices.

Let's look at one last example:

$$12 \times \frac{1}{1000} = 12{,}000 \text{ ?!}$$

Here the student probably pressed $\boxed{\times}$ instead of $\boxed{\div}$. Since 1000 is in the denominator, it will make 12 one thousand times smaller instead of one thousand times larger.

Now try the practice problems on page 169.

Get to Know Your Textbook

Don't be overwhelmed by the size of the text! Look at your syllabus and see how much of the text you are actually going to cover during the term. Usually it is less, sometimes much less, than the entire book.

Read all of the prefaces and introductions, especially those labeled "To the Professor" or "To the Student." By reading these sections you will get a feel for the real human being(s) who wrote the book and for their style of writing. The author(s) will also explain how they approached chemistry, how they organized the text, what they think is special about their book, and what study aids they have included.

Look at the Table of Contents to see what is included in the text. In addition to the content chapters, most texts also contain:

- prefaces and/or introductory material, as mentioned above.
- a glossary.
- an index.
- an answer key for at least some of the practice problems.
- appendices containing math reviews, conversion tables, formulas, etc.

Look inside the front and back covers. Usually you'll find the periodic table, an alphabetical listing of the elements with their atomic numbers and atomic weights, and perhaps other reference materials.

Look at how the chapters are organized. Most textbook chapters will have:

- an introduction that will state the goals of the chapter and what previous information is important for understanding the material in that chapter.
- section headings that point out the main ideas that are covered in the chapter.
- a summary at the end of the chapter.

- a list of keywords at the beginning or the end of the chapter.

- practice problems within the chapter.

- practice problems at the end of the chapter.

- relevant pictures and diagrams.

- marginal notes which summarize key definitions and concepts.

- pages that are often brightly colored and titled "connections," "extensions," "interesting applications," etc. Don't skip these sections because they show how the material you are learning is actually related to real life.

CHAPTER 4

Rulers, Weights, and Measuring Cups
Measurement and the Metric System

Chemistry is a quantitative science. You can look at the world two ways: qualitatively or quantitatively. The redness of an apple, the stickiness of an adhesive, the smell of a perfume, or even your personality are qualitative properties. Your height, your weight, your bank balance, or the amount of gasoline you buy are quantities. Sometimes there is a fine line between qualities and quantities. The redness of an apple can be measured as a wavelength of so many nanometers of light. The stickiness of an adhesive can be measured in the pounds of force it takes to peel it off. And psychologists even try to quantify something as indefinable as a personality with various psychological tests. On the other hand, for some people, even their bank balance is qualitative. Instead of an exact balance they say, "I've got some," or "I've got enough to cover that check."

You may be a little dismayed that chemistry often deals with quantities. But remember, you work with quantities constantly in your everyday life. You buy 10 gallons of gasoline. You add 1/2 teaspoon of salt to a recipe for bread. You weigh yourself in pounds on your bathroom scale.

Gallons, teaspoons, and pounds are examples of units that you use in everyday life. Units are standards of measurement we all agree upon so that we talk the same language. If the author of a recipe tells you to add 2 teaspoons of salt, you, she, and the teaspoon manufacturer have to agree on the definition of a teaspoon. If there were no agreement on units like teaspoons, it would be difficult to trade recipes!

Likewise, if we all did not agree on temperature, it would be hard to tell someone else how cold it is outside. What if the weatherman, instead telling us a temperature, just reported it was "hot," "cool," "cold," or "real, real cold"? What might be "cool" to him or her might be "real, real cold" to you. Fortunately most weather reporters give us the temperature as a quantity with units of degrees. If the weatherman says the temperature is 5°F, you can make your own judgment based on your own experience of 5°F temperature.

How Big are the King's Feet?
Measurement and the English System

Before we go into the metric system units you will use in chemistry, let's look at our own English units that we are more familiar with. In any system of measurement, there are three fundamental properties we measure: length, mass, and volume.

Length is the measurement of distance. We measure large distances, like the distance across town or the distance between cities, in miles. We measure medium size lengths such as the size of buildings and rooms in feet and yards, And we measure small distances, like the size of postcards, in inches.

As you may know, these units are very ancient. A "foot" was decreed to be the size of the King's foot. (The king that gave us our modern foot must have had quite big feet!) An inch was the size of a knuckle. Henry I of England decreed that a yard was to be the distance from his nose to his thumb when his arm was outstretched.

The next fundamental unit is mass. Mass and weight are often confused with each other in everyday language. Mass is an object's resistance to change in motion. A semi-truck has much more mass than a little Volkswagen. It's hard to get a semi going when it's stopped (as we all have experienced as we've sat behind one after the light has turned green). Similarly it is very hard to stop a semi once it's going fast.

Weight, on the other hand, is the gravitational attraction for a mass. Thus if you weigh 180 pounds on earth, you will weigh 30 pounds on the moon, which has 1/6 the gravity of earth. And in outer space you will be weightless. You will be weightless but not massless. If you get clanked on the head in outer space by a weightless hammer, you would still feel it. It is the mass, not the weight, that you feel.

The pound is actually a weight unit, not a mass unit. However, in common usage, we do not make a distinction between weight and mass. The distinction is not important unless we are trying to use a spring scale, which depends on gravity, or we are on the moon or in the space shuttle.

The last fundamental property is volume. Volume is the measurement of the amount of space something occupies. Your body takes up space. You can see this the next time you get in the bathtub and watch the water level as your body displaces some of the water.

We have many English units for volume: gallons, quarts, pints, cups, tablespoons, teaspoons, fluid ounces. We buy milk and ice cream in gallons, quarts, and pints. We measure recipes in cups, tablespoons, and teaspoons and buy medicine by the fluid ounce.

Another way to think of volume is length cubed. We live in three-dimensional space: an object has length, width, and depth. If we have a box that is 1 foot high, 1 foot wide, and 1 foot deep, the box has a volume of 1 $foot^3$, which we read as one cubic foot. An aquarium that is 2 feet wide, 1 foot high and 1 foot deep has a volume of 2 $feet^3$. Volumes of refrigerators and swimming pools are measured in cubic feet. If you use natural gas, look on your bill and you will see that gas usage is reported in cubic feet.

Oh, Those French—The Metric System

The problem with our English system, as you might already be aware, is that it is very confusing to convert between units. Imagine if you came here from a foreign country and had to memorize 5,280 feet to a mile, 4 quarts to a gallon, 16 ounces to a pound, 2000 pounds to a ton, etc. You would go crazy!

That is just what some French revolutionaries thought in the 1790s, during the French Revolution. They developed a system of measurement we now call the "metric system." Their idea was to base their new system of units on powers of ten, just like our number system. They reasoned that it would be easy to use and remember. They did not want to base length on a king's arms or feet, since they had overthrown the King, so they defined the metric unit of length, the meter, as 1/10,000,000 of the distance from the equator to the north pole (on the line that goes through Paris, of course). This length worked out to be a little longer than a yard.

They went on and did the same thing with mass, volume and temperature units, as we shall soon see.

Well, the ball started rolling in France, and now every country in the world, including England, is metric with one exception—the USA! Actually our government officially adopted the metric system several times, first in 1866 and then again in 1974 and 1985. But except for scientific use, the metric system never caught on. If it had, and we were raised with the metric system, you would not need to read this chapter.

So, the problem most Americans have with the metric system is that it is not familiar. It does not come naturally to us like our inches, quarts, and pounds that we know so well. What we want to do in the rest of this chapter is to help make the metric system familiar for you. If working metric problems is not just manipulating numbers (as it is for most people in beginning chemistry classes), you will make far fewer mistakes. It will also be easier to learn, and you will remember it much longer.

Give Them a Centimeter and They'll Take a Kilometer

The fundamental metric unit of length is the meter, which is a little longer than a yard (39.37 inches to be exact). For measuring small distances we need to have subdivisions of the meter. In the English system, an inch is a subdivision of the foot. Subdivisions are handled elegantly in the metric system with prefixes. The neat thing is that the prefixes will be the same no matter what units we attach to them. Once you learn them for one unit, they work the same way for any other unit. Table 4.1 lists the metric prefixes. Let's see how the prefixes work with the meter.

deci—the prefix deci means $\frac{1}{10}$ or 0.1. The Latin root "deci" is where we get our base 10 "decimal" system.

If we put the prefix deci in front of meter we get

$$1 \text{ decimeter} = 0.1 \text{ meter}$$

If a decimeter is 0.1 (one tenth) of a meter, then we can also say there are 10 decimeters (dm) in a meter.

$$10 \text{ dm} = 1 \text{ m}$$

Figure 4.1 shows these two relationships.

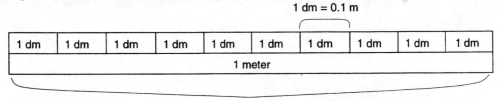

Figure 4.1

A decimeter is about the length of a finger.

centi—the prefix centi means $\frac{1}{100}$ or 0.01.

"centi" comes from the same Latin root as cent which is 0.01 of a dollar.

If we put the prefix centi in front of meter we get

$$1 \text{ centimeter} = 0.01 \text{ meter}$$

If a centimeter is 0.01 (one hundredth) of a meter, then we can also say there are 100 centimeters in a meter.

$$100 \text{ cm} = 1 \text{ m}$$

A centimeter is about the width of a fingernail.

🙣   **milli**—the prefix milli means $\frac{1}{1000}$ or 0.001

"milli" comes from the same root as *mill* in property taxes. A mill is 0.001 of a dollar.

If we put the prefix milli (abbreviated m) in front of meter we get

$$1 \text{ mm} = 0.001 \text{ meter}$$

If a millimeter is 0.001 (one thousandth) of a meter, we can also say there are 1000 millimeters (mm) in one meter.

$$1000 \text{ mm} = 1 \text{ m}$$

A millimeter is about the thickness of a dime.

🙣   **micro**—the prefix micro means one millionth, which we can write as

$$\frac{1}{1,000,000} \text{ or } \frac{1}{10^6} \text{ or } 10^{-6}.$$

When the numbers get this small, you can see it is easier to put them in exponential form.

If we put the prefix micro (abbreviated μ) in front of meter we get

$$1 \mu\text{m} = 10^{-6} \text{ meter}$$

or, written with a positive exponent

$$10^6 \mu\text{m} = 1\text{m}$$

Since a micrometer is one millionth (10^{-6}) of a meter, there are a million micrometers (10^6) in a meter. The cells of your body are about 3 to 5 μm in diameter.

nano—nano means one billionth, which we can write as $\frac{1}{1,000,000,000}$ or $\frac{1}{10^9}$ or 10^{-9}. So if we put the prefix nano (abbreviated n) we get:

$$1 \text{ nanometer} = 10^{-9} \text{ meter}$$

or in whole numbers

$$10^9 \text{ nm} = 1\text{m}$$

A nanometer is a billionth of a meter, therefore there are a billion nanometers in a meter. A nanometer is about the size of a molecule in your body.

Table 4.1: The Most common Metric Prefixes

Prefix	Symbol	Base Unit Multiplied By
kilo	k	1000 or 10^3
deci	d	0.1 or 10^{-1}
centi	c	0.01 or 10^{-2}
milli	m	0.001 or 10^{-3}
micro	μ	0.000001 or 10^{-6}
nano	n	0.000000001 or 10^{-9}

Keep in mind with all these prefixes we are taking a meter and cutting it into smaller and smaller pieces: 10 pieces for the decimeter, 100 pieces for the centimeter, 1,000 pieces for the millimeter, 1,000,000 pieces for the micrometer, and 1,000,000,000 pieces for the nanometer.

Besides subdivisions, we can also have multiples of the basic unit. In the English system the mile is a multiple of the foot (1 mile = 5,280 feet). In chemistry the only multiple we commonly use is kilo.

$$kilo = 1000$$

So

$$1 \text{ kilometer} = 1000 \text{ meters}$$

If you lay 1,000 meter sticks end-to-end, you get a kilometer, which is just a little over six tenths of a mile. To get a feeling for how long a kilometer is, the next time you drive from your house, notice when the odometer moves 0.6 miles. That distance is about a kilometer.

Students often mix up millimeters and kilometers when they do problems. If you make the metric system concrete, this will not happen to you. A millimeter is the thickness of a dime. If you stacked 1000 dimes, you would get about a meter (1000 mm = 1 m) If you lay a thousand meter sticks (each a little longer than a yard) end to end, you get 1 kilometer (1 km = 1000 m).

A Gram of Prevention is Worth a Kilogram of Cure

The same prefixes we used for the meter can be used for the gram. However, in chemistry decigrams and centigrams are rarely used. But we often use:

$$1 \text{ milligram} = 0.001 \text{ gram}$$

or in whole number form

$$1000 \text{ mg} = 1 \text{ g}$$

$$1 \text{ microgram} = 10^{-6} \text{ gram}$$

or written with a positive exponent

$$10^6 \mu g = 1 \text{ g}$$

$$1 \text{ nanogram} = 10^{-9} \text{ gram}$$

or written in whole number form

$$10^9 \text{ ng} = 1g$$

A gram is about 1/28 of an ounce or about the weight of an aspirin pill. A milligram is about the weight of a gnat or a pinch of salt.

We can also use the prefix kilo with the gram

$$1 \text{ kg} = 1000 \text{ g}$$

A kilogram is about two pounds (2.2 pounds to be exact).

Just a Milliliter of Sugar Helps the Medicine Go Down

The metric unit of volume is the liter. The liter is defined as a cube that is 10 cm on each side. It is 10 cm wide, 10 cm high and 10 cm deep. We multiply

$$10 \text{ cm} \times 10 \text{ cm} \times 10 \text{ cm} = 10^3 \; 10^3 = 1000 \text{ cm}^3$$

$$10 \times 10 \times 10 \times \text{cm}^3 \times \text{cm}^3 \times \text{cm}^3 = 10^3 \text{ cm}^3 = 1000 \text{ cm}^3$$

Units multiply and divide just as numbers do. So $10 \times 10 \times 10 = 10^3$ and $\text{cm} \times \text{cm} \times \text{cm} = \text{cm}^3$. The big one liter cube has 1000 little centimeter cubes in it.

If you fill a one liter cube with water and then pour the water into a bottle, you still have one liter. The shape of the liter does not matter. We defined a liter as a cube, but now we can forget about the cube and just remember the conversion:

$$1 \text{ Liter} = 1000 \text{ cm}^3$$

A liter is just a little bigger than a quart. A liter bottle of soda gives you about 6% more than a quart.

We can use the metric prefixes to make subdivisions of the liter.

$$1 \text{ deciliter} = 0.1 \text{ liter}$$

or using whole numbers

$$10 \text{ dL} = 1 \text{ L}$$

Deciliters are not used very often in chemistry, but they are often used in hospitals where blood test results are reported in milligrams per deciliter. A deciliter is about a half cup.

The primary subdivision of the liter we use in chemistry is the milliliter.

$$1 \text{ milliliter} = 0.001 \text{ liter}$$

or using whole numbers

$$1 \text{ L} = 1000 \text{ mL}$$

Since we also know that

$$1 \text{ L} = 1000 \text{ cm}^3$$

we conclude

$$1 \text{ mL} = 1 \text{ cm}^3$$

A milliliter is exactly the same size as a cubic centimeter. This information will be important later in density problems where milliliters and cubic centimeters are used interchangeably. In hospitals, cubic centimeters are often called "cc's". So we can say

$$1 \text{ mL} = 1 \text{ cm}^3 = 1 \text{ cc}$$

A sugar cube about the width of your fingernail is about the size of a milliliter. A teaspoon is about 5 milliliters.

Other Metric Volumes

You may have noticed we did not use the metric prefix kilo with liters. It is never used. Instead for volumes larger than the liter the cubic meter is used. A cubic meter is a cube that is 1 meter on each side. Imagine a box that is a little longer than a yard on each side. A cubic meter holds 1000 liters.

$$1 \text{ m}^3 = 1000 \text{ L}$$

So a cubic meter is really a "kiloliter." We do not use the term "kiloliter" because the meter is considered a more fundamental unit than the liter. You use cubic meters for large volumes like swimming pools or large tanks. Many natural gas companies now use cubic meters instead of cubic feet for billing.

Sometimes You Have to Sigh
A Note on SI Units

In 1960 scientists revised and further standardized the metric system to give it more uniformity. They called the revised system The International System of Units or SI (after the French Systéme Internationale). The basic SI unit of mass is the kilogram, rather than the gram as it is in the metric system. Likewise the SI unit of volume is the cubic meter (m^3) rather than the liter. The meter is the basic unit of length in both systems.

In most cases you will use metric units. But occasionally you may have to use SI units. When you study the properties of gases you will use the SI unit of temperature, the Kelvin (K), rather than the metric unit of Celsius (°C). And when you study heat, you will learn both the SI unit, the Joule (J), and the metric unit, the calorie (cal).

Reading the Want Ads
Kilodollars and other Units

The metric prefixes can be used with any units. If you read the want ads, you might read salary "28-35K$"or simply "28-35K." These ads are using "kilodollars."

$$1 \text{ kilodollar} = 1000 \text{ dollars}$$

So 28 to 35 kilodollars is $28,000- $35,000.

It doesn't matter if you know what the unit means. For instance, you know from their prefixes that

$$1 \text{ kilojoule} = 1,000 \text{ joules}$$

$$100 \text{ centipoise} = 1 \text{ poise}$$

Table 4.2 lists real things the size of each metric unit. Study this table and try to think of your own examples. After a while you will begin to make metric measurements concrete and real so that you will not just be memorizing numbers. Then you will really start thinking like a European and making conversions will become a simple matter. Table 4.3 lists in one place all the metric conversions you will need to memorize. The next chapter will show you a very easy method to convert between metric units and to convert metric units to English units.

Table 4.2: Metric Sizes of Everyday Objects

Length		Mass	
meter	a yardstick, about waist high	gram	two aspirin tablets
decimeter	length of finger	milligram	a gnat
centimeter	width of fingernail	kilogram	a little over 2 lbs of cheese
millimeter	thickness of a dime	**Volume**	
kilometer	a little over half a mile	liter	a quart
		deciliter	half of a cup
		milliliter	1/5 of a teaspoon

Table 4.3: The Most Common Metric Conversions

Length	Mass	Volume
1 m = 10 dm	$1\ g = 1000\ mg = 10^3\ mg$	1 L = 10 dL
1 m = 100 cm	$1\ g = 10^6\ \mu g$	$1\ L = 1000\ mL = 10^3\ mL$
$1\ m = 1000\ mm = 10^3\ mm$	$1\ g = 10^9\ ng$	$1\ mL = 1\ cm^3 = 1cc$
$1\ m = 10^6\ \mu m$	$1\ kg = 1000\ g = 10^3\ g$	$1\ m^3 = 1000\ L = 10^6 cm^3$
$1\ m = 10^9\ nm$		
$1\ km = 1000\ m = 10^3\ m$		

How to Take Notes in Class

Sit in the front and middle of the class, especially in large and noisy lecture halls. It will be easier to hear and see. It will also force you to stay alert and take notes because you will be right in the line of sight of the instructor!

Get a large notebook with pockets or a looseleaf notebook so you will have a place to put the syllabus and any other handouts. Don't be afraid to use paper. One of the worst things you can do is to write your notes small and all scrunched together. You will go blind trying to decipher them later, and you will have no room to add extra information or to clarify something.

Prepare for the lecture before class by:

- reviewing the notes from the previous class.

- reviewing the homework.

- glancing over the chapter sections that will be covered that day. Don't worry if you don't understand everything. At least the instructor's terms will be familiar.

During class:

- Write on the right hand two-thirds of the page. Save the left hand side for notes when you review after class and for writing key words that will help you when studying for a quiz or test. (Study skills texts refer to this method as the Cornell Method of notetaking.)

- Write the date at the beginning of each day's notes.

- Take notes as completely as you can while still understanding.

- Write down each step of every problem even if you do not understand the step. You can always ask about it later.

- Write a question mark next to anything you don't understand. Come back to it later. Try to find the topic in the text or get an explanation from the professor, a tutor, or better yet, from a classmate.

- Don't just listen. Be an ACTIVE LEARNER. Nod when you understand something. Give yourself (and the professor) some positive reinforcement!

- Listen for clues such as, "This definition is very important," or "I don't have time to cover this in class, but it is required," or "This is not covered in your text-book."

🍃   Use shorthand and abbreviations so that you can write quickly but understandably. Examples are BP for boiling point and SH for specific heat. The first time you use an abbreviation write its translation, then continue with just the abbreviation.

🍃   Don't get lost in the five minute rustle. Five or ten minutes before class ends, some students start rustling their books, papers and bookbags in anticipation of running off to their next exciting activity. This is especially a problem in large lecture halls. Don't neglect these last minutes of class; you can miss some important information. Being able to hear clearly during the five minute rustle is also a good reason for sitting in the front and center.

As soon as possible after class:

🍃   Review your notes while they are still fresh in your mind. This will help to move the information from your short-term memory to your long-term memory. (If at all possible, never, never schedule a class immediately after any course that you feel is going to be difficult for you. You need the hour immediately after that class to go over notes and begin homework while the material is still fresh. Studies tell us that if you go over material within an hour or so after class, you will remember about 80% of what you have learned. If you wait until later, you will forget 80% of what you have learned. That's a big difference!)

🍃   Mark your notes. Highlight with a marker. Write notes in the left-hand margin. Take ownership of your notes and your text.

🍃   Above all else never, never miss a lecture! Chemistry is cumulative. What is presented tomorrow will depend upon your knowledge of what was covered today. If for any reason you have no choice other than to miss a class, arrange to get someone's notes and go over them before the next lecture. (Be sure to get notes from someone who is as good a notetaker as you are!)

CHAPTER 5

It's the "One" Thing To Do
Unit Conversions

In the last chapter you looked at conversions between metric units. Those conversions are listed for you in one convenient place in Table 4.3. You also started to make the metric units more familiar by relating their sizes to everyday items and distances.

In this chapter we will show you a very easy way to convert between metric units and also between metric and English units. We call the process "unit conversions." It goes by other names in some texts such as "factor labeling," "unit analysis," or "dimensional analysis." The process is the same no matter what we call it.

Most of the math you will do on your first test will involve unit conversions. This is a powerful technique to learn. It is not just memorizing formulas. It is a way of thinking that you can use on your first test, on problems later in the class, and even to solve problems in everyday life and your career.

What Do Unit Conversions Mean to Me?
Doing Conversions

Before we look at the metric system, let's start with a English relationship we know.

12 inches = 1 foot

This statement is an equality, that is, we are saying 12 inches is exactly the same as 1 foot. If we multiply or divide both sides of an equality by the same thing, we do not change the relationship. It remains an equality. Let's divide both sides by 1 foot.

$$\frac{12 \text{ inches}}{1 \text{ foot}} = \frac{1 \text{ foot}}{1 \text{ foot}} = 1$$

Anything divided by itself equals one and units cancel just as numbers do. Therefore we have derived a fraction, $\frac{12 \text{ inch}}{1 \text{ foot}}$, that is equal to one.

Now let's divide the equality by 12 inches:

$$\frac{12 \text{ inches}}{12 \text{ inches}} = \frac{1 \text{ foot}}{12 \text{ inches}} = 1$$

Therefore from the equality

$$12 \text{ inch} = 1 \text{ foot}$$

we can derive two fractions that are equal to one.

$$\frac{12 \text{ inches}}{1 \text{ foot}} \quad \text{and} \quad \frac{1 \text{ foot}}{12 \text{ inches}}$$

Notice that one conversion is the reciprocal of the other, that is, it is the other turned upside down.

Now let's use these conversion factors to do some conversions between inches and feet. These conversions may seem very simple, but you will do exactly the same process when you convert between units in the metric system and when you convert between moles and grams later in the class.

Remember to look for patterns!

Let's convert 4.5 feet to inches.

$$4.5 \text{ feet} \times \frac{12 \text{ inch}}{1 \text{ foot}} = 54 \text{ inch}$$

Important Point

We multiplied by the conversion factor with the foot units on the bottom so they cancel, and we end up with the unit we are trying to solve for, inches. No matter what the units are, the pattern for converting will be the same:

$$\text{old units} \times \frac{\text{new units}}{\text{old units}} = \text{new units}$$

Now let's convert 30 inches to feet.

$$30 \text{ inch} \times \frac{1 \text{ foot}}{12 \text{ inch}} = 2.5 \text{ feet}$$

The old units, inches, cancel and we end up with the new units, feet.

Always label the numbers in a problem with units. Without units many mistakes can go undetected!

One of the great things about unit conversions is they are self-checking. What if we made this mistake:

$$30 \text{ inch} \times \frac{12 \text{ inch}}{1 \text{ foot}} = \text{?!}$$

You can immediately see that our units do not cancel. So you know right away that the conversion needs to be inverted. Using unit conversions makes it almost impossible to make a mistake.

But Where Do You Find Conversions?

You have used some conversions so often in your life you just know them. These are conversions like 12 inches to 1 foot and other English system conversions such as 16 fluid ounces to 1 pint or 4 quarts to 1 gallon. You may need to look up other conversions. For a while you will need to look up the metric conversions in Table 4.3. We hope that these conversions will also become second nature to you soon. But for now, don't be ashamed to look them up.

Now let's try some metric conversions. Let's convert 2.57 meters to millimeters. Before we do the conversion, look up the conversion between millimeters and meters in Table 4.3. It is

$$1000 \text{ mm} = 1 \text{ m}$$

You can write the two conversions we get from this equality:

$$\frac{1000 \text{ mm}}{1 \text{ m}} \quad \text{or} \quad \frac{1 \text{ m}}{1000 \text{ mm}}$$

Now let's do the conversion. First we write down the given quantity, then multiply by a conversion factor.

$$2.57 \text{ m} \times \frac{1000 \text{ mm}}{1 \text{ m}} = 2570 \text{ mm}$$

We chose the first conversion with meters in the denominator so that they cancel the meter units in the given quantity. On the calculator:

Press 2.57 $\boxed{\times}$ 1000 $\boxed{=}$ The display reads $2570.$

Notice that since the 1000 is in the numerator, we multiply by 1000 and the decimal point moves three places to the right, that is, it gets larger by 1000.

Now let's convert 435 mm to m.

We use the same conversion factor 1000 mm = 1 m

$$435 \; \cancel{\text{mm}} \times \frac{1 \text{ m}}{1000 \; \cancel{\text{mm}}} = 0.435 \text{ m}$$

Since the 1000 is in the denominator, we divide. On the calculator:

Press 435 $\boxed{\div}$ 1000 $\boxed{=}$ The display reads 0.435

If we use the wrong conversion, as shown below, the units will not cancel and we will see that we have to turn the conversion upside down.

$$435 \text{ mm} \times \frac{1000 \text{ mm}}{1 \text{ m}} = ?!$$

Beware! There is one way you can make a mistake that will be hard to catch using unit conversions. That is if you memorized the metric conversion between meters and millimeters **incorrectly** as

$$1 \text{ mm} = 1000 \text{ m} \quad \text{(incorrect!)}$$

If you did this, you would set up the problem in a way which looked correct to you, because the units would cancel. But it would be wrong because you were using the incorrect numbers in the conversion. Let's show what would happen if you used the incorrect metric conversion to convert 1.662 meters to millimeters:

$$1.662 \; \cancel{\text{m}} \times \frac{1 \text{ mm}}{1000 \; \cancel{\text{m}}} = 0.001662 \text{ mm} ?!! \quad \text{(incorrect!)}$$

Look carefully at the incorrect answer. Does it make sense? If you have made the metric system concrete, you will recall that a millimeter is only about the thickness of a dime, while a meter is about a yard. So your answer is saying that something that is about 1¾ yards long is about 1/1000 the thickness of a dime. Of course this comparison could not be correct! Now look at the conversion you wrote and you will realize that you are saying the thickness of a dime is the same as a thousand meter sticks! You then catch your mistake and write the conversion correctly as $\dfrac{1000 \text{ mm}}{1 \text{ m}}$ and redo the problem:

$$1.662 \text{ m} \times \frac{1000 \text{ mm}}{1 \text{ m}} = 1662 \text{ mm}$$

Remember, to a person from France, 1 mm equals 1000 m would be as immediately absurd as 12 feet equals 1 inch is to you.

How Do I Move the Decimal Point?
Metric Conversions

Now let's do a few more metric conversions. Again, all the conversions are found in Table 4.3.

Convert 27 cm to m.

$$27 \text{ cm} \times \frac{1 \text{ m}}{100 \text{ cm}} = 0.27 \text{ cm}$$

Convert 4.21 kg to grams.

$$4.21 \text{ kg} \times \frac{1000 \text{ g}}{1 \text{ kg}} = 4210 \text{ g}$$

Convert 35.0 mL to liters.

$$35.0 \text{ mL} \times \frac{1 \text{ L}}{1000 \text{ mL}} = 0.0350 \text{ L}$$

Note on Significant Figures

Exact numbers, such as those in conversion factors, do not affect the number of significant figures in a calculation. There are exactly 100 centimeters in one meter. There are exactly 1000 grams in one kilogram and 1000 milliliters in one liter. So in the three examples above, the answers have the same number of significant figures as the starting quantities. Likewise in the English system there are exactly 12 inches to a foot, 4 quarts to a gallon, 16 ounces to a pound, etc.

Multiple Conversions

Sometimes you may have to do more than one conversion in a problem. This often happens when neither of the units is the basic unit.

Convert 0.25 km to cm.

First we convert km to meters, the basic unit, then we convert meters to cm.

That is, we go km ⇨ m ⇨ cm.

$$0.25 \text{ km} \times \frac{1000 \text{ m}}{1 \text{ km}} \times \frac{100 \text{ cm}}{1 \text{ m}} = 25000 \text{ cm}$$

Or in scientific notation: 2.5×10^4 cm

Now the nice thing about using the calculator is that you can line up all the conversions in one problem. On your calculator:

Enter 0.25 [×] 1,000 [×] 100 [=]   The display reads $25000.$

You can then press [MODE] [8] [8] to convert to scientific notation if you choose. The answer then becomes 2.5×10^4 cm.

Then press [MODE] [9] to convert back to decimal form. Otherwise your calculator will remain in scientific notation—as we warned you in Chapter 3.

Let's try another. Convert 7.44×10^4 cm to kilometers.

The conversion will go like this: cm ⇨ m ⇨ km.

$$7.44 \times 10^4 \text{ cm} \times \frac{1 \text{ m}}{100 \text{ cm}} \times \frac{1 \text{ km}}{1000 \text{ m}} = 0.744 \text{ km}$$

Press 7.44 [EE] 4 [÷] 100 [÷] 1000 [=] The display reads 0.744

Press [MODE] [8] [8] to convert to scientific notation 7.44×10^{-1}

Chapter 5

Switching Systems
Metric-English Conversions

What if you want to convert between the English system and the metric system? Table 5.1 lists some common conversions between the two systems. Your textbook will have a similar table. Your instructor may want you to memorize a few conversions, such as 1.61 km to 1 mile or 1.06 qt to 1 L, but you probably won't be asked to memorize them all. However, you do need to know where to find them and how to use them.

Table 5.1 Metric-English Conversions

Length	Mass	Volume
1 m = 39.4 in.	1 kg = 2.20 lbs.	1 L = 1.06 qt
2.54 cm = 1 in.	454 g = 1 lb.	1 mL = 5 tsp
1.61 km = 1 mile		946 mL = 1 L

These conversions work the same as any other unit conversions. Let's try some.

How many miles is a 10.0 kilometer race (often called simply a "10 K")?

We look up the conversion between miles and kilometers in Table 5.1 and find

$$1.61 \text{ km} = 1 \text{ mile}$$

So lining up the conversions:

$$10.0 \text{ km} \times \frac{1 \text{ mile}}{1.61 \text{ km}} = 6.21 \text{ miles}$$

We round the answer off to 3 significant figures.

A piece of cloth is 47 inches long. How many centimeters long is it?

Look up the conversion between cm and inches in Table 5.1. We find it is

$$2.54 \text{ cm} = 1 \text{ in.}$$

So can write the conversion so that the inch units cancel

$$47 \text{ inch} \times \frac{2.54 \text{ cm}}{1 \text{ inch}} = 120 \text{ cm}$$

Press 47 [×] 2.54 [=]   The display reads   *119.38*

We round the answer off to 2 significant figures to 120 cm, or in scientific notation to clearly show that we have two significant figures 1.2×10^2 cm.

What if you want to order 1.5 lbs of cheese in Canada where they use the metric system? How many kilograms do you order? Look up the conversion.

$$1.5 \; \cancel{lb} \times \frac{1 \; kg}{2.2 \; \cancel{lb}} = 0.68 \; kg$$

Sometimes we need to do more than one conversion depending on what conversions we find in the table. For example, suppose we want to convert 0.500 gallons to liters. When we look up the conversion in Table 5.1, we only find

$$1.06 \; qt = 1 \; L$$

Even though we do not find a conversion between gallons and liters, we can still do the problem because we know the English relationship

$$4 \; qts = 1 \; gal$$

So we can use gallons ⇨ quarts ⇨ liters.

$$0.500 \; \cancel{gal} \times \frac{4 \; \cancel{qt}}{1 \; \cancel{gal}} \times \frac{1 \; L}{1.06 \; \cancel{qt}} = 1.89 \; L$$

Press .5 $\boxed{\times}$ 4 $\boxed{\div}$ 1.06 $\boxed{=}$ The display reads   *1.886792453*

We round the answer to 3 significant figures.

Reading the Tables
Where Do You find Conversions?

There are three ways you can find conversions. First, conversions are often listed in tables like the metric conversions in Table 4.3 or the metric-English conversions in Table 5.1 in this book. You will find similar tables in your textbook. Second, there are many conversions you know because you have worked with them so often in everyday life. You are probably thoroughly acquainted with English conversions such as:

$$5{,}280 \; feet = 1 \; mile$$
$$100 \; cents = 1 \; dollar$$
$$2 \; pints = 1 \; qt$$
$$4 \; quarts = 1 \; gallon$$
$$16 \; ounces = 1 \; pound$$

A third source of conversions is to figure them for yourself. Anytime you see the word "per" you have a conversion because "per" means to divide. For example, if a typist types 75 words per minute, you can write this conversion: $\dfrac{75 \text{ words}}{1 \text{ min}}$. If you earn $8.00 per hour, you earn $8.00 in one hour. The conversion, then, is $\dfrac{8 \text{ dollars}}{1 \text{ hr}}$.

Question: If you earn $8.00 per hour and work 78 hours in two weeks, how much will you earn?

$$78 \text{ hours} \times \frac{8 \text{ dollars}}{1 \text{ hour}} = \$624$$

If you want to have $300 for a new TV set, how many hours will you have to work (without taking into account taxes)?

$$\$300 \times \frac{1 \text{ hour}}{\$8.00} = 37.5 \text{ hours}$$

We turn the conversion $8.00 per hour upside down so the "$" units cancel.

Looking Ahead

Later in the class, in the chapter on moles, you will see conversions that look like this:

16 grams of methane = 1 mole of methane

or like this

180 g of glucose = 1 mole of glucose

You can use these conversions to convert back and forth between moles and grams of these substances. You don't need to know what a mole is or even what a gram is right now to do these conversions. You just need to know how to cancel units.

If there are 180.0 grams to one mole of glucose, how many grams are there in 1.52 moles?

$$1.52 \text{ mole} \times \frac{180.0 \text{ g}}{1 \text{ mole}} = 274 \text{ gram}$$

The calculator answer is 273.6, but 1.52 mole has three significant figures. Therefore we round our answer to three significant figures.

How many moles are there if you weigh out 125.0 grams of glucose?

$$125.0\,g \times \frac{1\ mole}{180.0\ g} = 0.6944\ moles$$

You can see that the math in this more advanced chapter is not any more difficult than what you are doing now.

Looking Ahead

Also in the chapter on moles you will learn that a "mole" has Avogadro's number of particles in it, which is 6.02×10^{23} particles. A mole of water has 6.02×10^{23} water molecules. A mole of carbon has 6.02×10^{23} atoms. A mole of anything is always going to have 6.02×10^{23} of whatever particles it is composed. That makes for easy conversions. Let's try some conversions and see that it's not that hard.

An eight ounce glass of water is about 12.5 moles. How many molecules does it contain? We use 1 mole water = 6.02×10^{23} molecules to do the conversion.

$$12.5\ \text{mole} \times \frac{6.02 \times 10^{23}\ \text{molecules}}{1\ \text{mole}} = 7.53 \times 10^{24}\ \text{molecules}$$

How many moles are there in one billion molecules of water (1×10^9 molecules)?

$$1 \times 10^9\ \text{molecules} \times \frac{1\ mole}{6.02 \times 10^{23}\ \text{molecules}} = 1.66 \times 10^{-15}\ \text{mole water}$$

Even a billion molecules makes a very small fraction of a mole.

Now try the practice problems on page 171.

Reading Your Textbook

If your assignment is to read pages 88-108, the worst thing you can possibly do, aside from not doing the reading at all, is to open the text to page 88 and begin reading. Reading an assignment is a three-phase process. It has a BEFORE, DURING, AND AFTER phase and each is equally important!

During the BEFORE phase:

- Skim the entire assignment first so you have a general idea of the content and try to relate the content to things you already know.

- Look at your syllabus and see how many lectures are devoted to the assignment. See what other material is included with this assignment on the next test.

- Judging from the length and the difficulty of the reading assignment, try to determine about how long the task will take and plan accordingly. (Remember this is technical reading, so reading only 1-2 pages per hour is not unreasonable.)

During the DURING phase:

- Mark up your text. Write notes in the margin. Highlight key concepts. (Try not to buy a book that has been marked up too much. You will be distracted by someone else's markings.)

- If examples do not show all the steps, write in the missing steps so you will remember how to do the problem when it is time to review for a quiz or test.

- Alice in Wonderland asked, "What good is a book without pictures?" Fortunately most modern chemistry books are filled with pictures. Take advantage of them. Remember, chemistry is visual. When the text refers to a drawing, figure, diagram, chart, photograph, etc., look at it right away. Study it. Sometimes one picture can be worth many words!

- Spend time reflecting on what you read. Can you put the concepts into your own words?

- Do you need to outline, make a diagram or write a summary of the material?

- Do you need to hear yourself reading the material? Reading difficult material aloud slows you down and lets you hear as well as see. Two senses are better than one! Remember, you are reading for understanding.

During the AFTER phase:

- Go back to any sections you did not understand. If after going over them again, you still have concerns, put a question mark beside them and ask about them during the next class or help session or meeting with your study group.

- Go back over the entire reading assignment looking at main headings and the notes you made in the margin. Does this information remind you of the main ideas and content of each section? If not, add to it.

- Write a brief summary of what you read. Include any type of diagram that will help you remember the material. This summary, what you have underlined in the text, what you have written in the margin in the text, and the notes you have taken in class are the materials you need to use when you review daily and study for quizzes and tests.

CHAPTER 6

How Big Is That Dose?
Bigger and Better Conversions

In the last chapter you saw that you can use unit conversions to convert between units in the metric system without having to worry about how to move the decimal point. You also learned to convert between metric and English units and got a preview in the *Looking Ahead* sections of conversions involving moles that you will perform later in the class.

In this chapter we will look at conversions that are a little more complex, including a few drug dosage problems that you will see later in your career if you go into nursing. You'll see that they are not that difficult if you use the unit conversion technique.

Is It Polite to Say MPH to Someone From France?
Slightly More Complex Conversions

Speed, the miles per hour (MPH) you read on your speedometer, is a relationship that converts between distance and time, miles and hours. We can write 65 miles per hour as

$\frac{65 \text{ miles}}{1 \text{ hr}}$ or its reciprocal $\frac{1 \text{ hr}}{65 \text{ miles}}$. Remember that "per" stands for a fraction line.

How far do you go if you drive 4 hours at 65 MPH?

Our given quantity is 4 hours. The conversion is 65 miles/hr.

$$4.0 \text{ hours} \times \frac{65 \text{ miles}}{1 \text{ hour}} = 260 \text{ miles}$$

The distance between Cleveland and Chicago is 500 miles. How long will the trip take if you drive 65 MPH?

We turn the velocity upside down so that the mile units cancel:

$$500 \; \cancel{\text{miles}} \times \frac{1 \text{ hour}}{65 \; \cancel{\text{miles}}} = 7.7 \text{ hours}$$

We have other conversions involving driving in our everyday life. One useful conversion is gas mileage in miles per gallon (MPG).

Suppose your car gets 35 miles per gallon and your gas tank holds 12 gallons. How far can you travel on that tank of gas?

We can write the two conversions for 35 mpg:

$$\frac{35 \text{ miles}}{1 \text{ gal}} \quad \text{and} \quad \frac{1 \text{ gal}}{35 \text{ miles}}$$

Now we'll write down the given quantity and do the conversion:

$$12 \; \cancel{\text{gal}} \times \frac{35 \text{ miles}}{1 \; \cancel{\text{gal}}} = 420 \text{ miles}$$

If you travel 500 miles to Chicago and get 40 mpg on the road, how many gallons of gas will the trip take? If gasoline costs \$1.20 per gallon, how much will you spend for gas?

This problem really has two parts. You will often see problems like this in chemistry textbooks. We will answer the first question and then use that answer to solve the second question. The first step is to convert the miles to gallons:

$$500 \; \cancel{\text{miles}} \times \frac{1 \text{ gal}}{40 \; \cancel{\text{miles}}} = 12.5 \text{ gallons}$$

The second step is to convert the gallons to dollars:

$$12.5 \; \cancel{\text{gal}} \times \frac{\$1.20}{1 \; \cancel{\text{gal}}} = \$15.00$$

Now suppose you have a friend from France. If you said "40 mpg" or "65 mph," it probably wouldn't mean a thing to her, but km per liter or km per hour would.

Let's help her out and convert 40 mpg to km/L and 65 mph to km/hr.

First let's convert the 40 mpg. We write it correctly as 40 miles/1 gal. We want to convert 40 miles/gal to km/L. There are two conversions here: miles to kilometers and gallons to liters. We can do them in either order. We'll convert the distance first and then the volume. We find in Table 5.1 the conversion between kilometers and miles (1.61 km = 1 mile) and the conversion between quarts and liters (1.06 qt = 1L).

$$\frac{40 \text{ miles}}{1 \text{ gal}} \times \frac{1.61 \text{ km}}{1 \text{ mile}} \times \frac{1 \text{ gal}}{4 \text{ qt}} \times \frac{1.06 \text{ qt}}{1 \text{ L}} = \frac{17 \text{ km}}{1 \text{ L}}$$

Remember that units in the numerator will cancel units in the denominator no matter where they occur in the problem. It does not matter if there is another fraction in between the units.

Now let's convert 65 miles per hour to kilometers per hour

$$\frac{65 \text{ miles}}{1 \text{ hr}} \times \frac{1.61 \text{ km}}{1 \text{ mile}} = \frac{105 \text{ km}}{1 \text{ hr}}$$

Now let's do a conversion within the American System using conversions we know from everyday life. Let's convert 65 miles/hr to feet/sec. That is, if you are shooting down the freeway at 65 mph, how many feet are ticking by each second?

We have two conversions: miles to feet and hours to seconds.

$$\frac{65 \text{ miles}}{1 \text{ hour}} \times \frac{5280 \text{ feet}}{1 \text{ mile}} \times \frac{1 \text{ hour}}{60 \text{ min}} \times \frac{1 \text{ min}}{60 \text{ sec}} = \frac{95 \text{ feet}}{1 \text{ sec}}$$

So when you are traveling at 65 mph, every second almost 100 feet are scooting by!

Now to show you how many conversion factors you can set up at one time, let's do a conversion to see how many seconds old someone is. We will do it for someone 24 years old, but you can substitute your age or anyone's age and do the conversion. So the question is: how many seconds old is someone who is 24 years old?

We want to go from years to seconds. There are 365 days in a year; 24 hrs a day; 60 minutes in an hour and 60 seconds in a minute. Actually there are 365.25 days in a year, to five significant figures. That extra 0.25 of a day is why we have leap years every four years. So we start with our given quantity of 24 years and keep converting until we get to seconds.

$$24 \text{ years} \times \frac{365.25 \text{ days}}{1 \text{ year}} \times \frac{24 \text{ hours}}{1 \text{ day}} \times \frac{60 \text{ minutes}}{1 \text{ hour}} \times \frac{60 \text{ seconds}}{1 \text{ minute}} = 7.5738 \times 10^8 \text{ seconds}$$

On the calculator press 24 ⊗ 365.25 ⊗ 24 ⊗ 60 ⊗ 60 ⊜

The display reads 7.5738240^{08}

Press (MODE) (9)   The display reads 757382400

So at 24 years, you've lived over 750 million seconds!

In Chapter 4 we mentioned that one teaspoon is about 5 milliliters. Let's prove that this statement is true using conversions.

In this problem we want to convert 1 tsp to mL

$$1 \text{ tsp} \Rightarrow ? \text{ mL}$$

We need some conversions. Look in most cookbooks and you will find:

$$3 \text{ teaspoons} = 1 \text{ tablespoon}$$

$$16 \text{ tablespoons} = 1 \text{ cup}$$

$$4 \text{ cups} = 1 \text{ quart}$$

In Table 5.1 we find the metric to English conversion:

$$1.06 \text{ qt} = 1 \text{ liter}$$

And you have memorized the metric conversion (if you haven't yet, you can find it in Table 4.3):

$$1000 \text{ mL} = 1 \text{ L}$$

Now write down the given quantity (1 tsp) and then line up the conversions so their units cancel:

$$1 \text{ tsp} \times \frac{1 \text{ Tsp}}{3 \text{ tsp}} \times \frac{1 \text{ cup}}{16 \text{ Tsp}} \times \frac{1 \text{ qt}}{4 \text{ cups}} \times \frac{1 \text{ L}}{1.06 \text{ qt}} \times \frac{1000 \text{ mL}}{1 \text{ L}} =$$

On the calculator we initially enter 1000, because it is the only number other than 1 in the numerators. We divide by the rest of the numbers since they are in the denominators:

1000 (÷) 3 (÷) 16 (÷) 4 (÷) 1.06 (=)   The display reads 4.913522

We round our answer to 1 significant figure to get 5 mL.

Will I Ever Use This After This Class?
Dosage Calculations

In nursing and in other classes that deal with medicines, you will often encounter dosage calculation problems. You will also encounter these types of problems on nursing board tests. The conversions in this section are probably more complex than any you will see in your class this semester. But if you can work your way through these problems, you can certainly handle any of the problems you will see in your class this term.

Using unit conversions makes doing dosage problems relatively easy. You can do even very complicated problems with just one line of conversion factors. And you can easily check the units by how they cancel. Let's try some and you'll see what we mean.

A patient weighs 175 pounds. He is to receive 0.5 mg of a medication per kilogram of body weight. The vial of the medicine is labeled 1.0 mg/2 mL. How many milliliters should be administered?

Before we solve let's look at each sentence. The first sentence tells us the patient weighs 175 pounds. This will be our starting quantity. The second sentence states that for every kilogram the patient weighs, he is to receive 0.5 mg of the medication. We can write this sentence as two conversions:

$$\frac{0.5 \text{ mg}}{1 \text{ kg}} \quad \text{and} \quad \frac{1 \text{ kg}}{0.5 \text{ mg}}$$

Also, the patient's weight is given in pounds, so we know we will have to convert his weight from pounds to kilograms. We find the conversion in Table 5.1.

$$2.2 \text{ lb} = 1 \text{ kg}$$

The third sentence says that the medicine's label states that it has 1.0 milligram of drug in every 2.0 milliliters of liquid. Most of the liquid in the medication is probably water and a little bit of alcohol (called inert ingredients). We can write this sentence as two conversions:

$$\frac{1.0 \text{ mg}}{2 \text{ mL}} \quad \text{and} \quad \frac{2 \text{ mL}}{1.0 \text{ mg}}$$

So we want to start with the patient's weight in pounds and work our way to milliliters of the medication solution. We set up the conversions so that the units cancel.

$$175 \cancel{\text{ lb}} \times \frac{1 \cancel{\text{ kg}}}{2.2 \cancel{\text{ lb}}} \times \frac{0.5 \cancel{\text{ mg}}}{1 \cancel{\text{ kg}}} \times \frac{2 \text{ mL}}{1.0 \cancel{\text{ mg}}} = 80 \text{ mL}$$

So this patient will get to drink 80 milliliters of the medicine.

Let's try another problem.

A physician orders 0.5 mg of epinephrine for a patient. The ampule containing the drug is labeled 0.1 mg/mL. How many milliliters of this drug should be given?

The first sentence gives us our starting quantity 0.5 mg of epinephrine. The next sentence gives us the conversion written on the label. Each milliliter contains 0.1 mg of epinephrine. We can write both conversion factors:

$$\frac{0.1 \text{ mg}}{1 \text{ mL}} \quad \text{and} \quad \frac{1 \text{ mL}}{0.1 \text{ mg}}$$

Now we will do the conversion.

$$0.5 \text{ mg} \times \frac{1 \text{ mL}}{0.1 \text{ mg}} = 5 \text{ mL}$$

Notice we used the second form of the conversion so that the mg units cancel.

Here is another problem.

A patient is to receive 1.0 gram of Vitamin C. The available Vitamin C tablets contain 250 mg per tablet. How many tablets should the patient be given?

The starting quantity is 1.0 gram. The second sentence gives us a conversion. We can write both forms of the conversion:

$$\frac{1 \text{ tablet}}{250 \text{ mg}} \quad \text{and} \quad \frac{250 \text{ mg}}{1 \text{ tablet}}$$

Since the conversion uses milligrams, we need to convert milligrams to grams.

$$1.0 \text{ g} \times \frac{1000 \text{ mg}}{1.0 \text{ g}} \times \frac{1 \text{ tablet}}{250 \text{ mg}} = 4 \text{ tablets}$$

The amount of fluid a patient receives from an intravenous (I.V.) bag is measured in drops. For instance, one particular I.V. device might be rated 15 drops = 1 mL. This means every 15 drops will measure 1 milliliter. This is a conversion and like any other conversion we can write the two forms:

$$\frac{15 \text{ drops}}{1 \text{ mL}} \quad \text{and} \quad \frac{1 \text{ mL}}{15 \text{ drops}}$$

The I.V. device can be set to measure out a given number of drops per minute. For instance, it can be set at 20 drops per minute. That is, if you took a watch and counted the drops in one minute, you would count 20. This can also be written as conversions in both forms:

$$\frac{20 \text{ drops}}{1 \text{ min}} \quad \text{and} \quad \frac{1 \text{ min}}{20 \text{ drops}}$$

Now let's try some problems that involve these conversions.

Intravenous lidocaine therapy is started for a patient. The doctor's order says to add 1.0 gram of lidocaine to 250 mL of I.V. solution and deliver it to the patient at 4.0 mg/min. In this particular I.V., 20 drops = 1 mL. What is the flow rate in drops per minute?

The second sentence gives two conversions. The first conversion says that there is 1 gram of lidocaine in 250 mL of solution. We can write the two forms of this conversion:

$$\frac{1.0 \text{ g}}{250 \text{ mL}} \quad \text{and} \quad \frac{250 \text{ mL}}{1.0 \text{ g}}$$

The second conversion states that the patient is supposed to receive 4.0 mg every minute $\frac{4.0 \text{ mg}}{1 \text{ min}}$. The third sentence gives a conversion between milliliters and drops.

$$\frac{20 \text{ drops}}{1 \text{ min}} \quad \text{and} \quad \frac{1 \text{ min}}{20 \text{ drops}}$$

Now we are going to convert the rate from $\frac{\text{mg}}{\text{min}}$ to $\frac{\text{drops}}{\text{min}}$.

$$\frac{4.0 \text{ mg}}{1 \text{ min}} \times \frac{1 \text{ g}}{1000 \text{ mg}} \times \frac{250 \text{ mL}}{1 \text{ g}} \times \frac{20 \text{ drops}}{1 \text{ mL}} = \frac{20 \text{ drops}}{1 \text{ min}} \text{ or 20 drops/min}$$

All the units cancel except for drops in the numerator and minutes in the denominator.

This problem is certainly more complex than any you'll typically encounter in your class, but you can see it's just a matter of finding the conversions in the problem and then lining them up so that the units cancel.

Try another example.

A patient is to receive 2400 mL of D5W solution (a typical I.V. solution of sugar in water) over 24 hours. How should the nurse set the flow rate on the I.V. in drops per minute if 15 drops equals 1 mL?

What we are given in the first sentence, 2400 mL per 24 hours, is a rate of volume of fluid per time. What we are being asked to convert to, drops per minute, is also a volume of fluid per time.

We want to convert $\dfrac{2400 \text{ mL}}{24 \text{ hr}}$ to $\dfrac{\text{drops}}{\text{min}}$

In the numerators we want to convert mL to drops. We are given that conversion in the second sentence, $\dfrac{15 \text{ drops}}{1 \text{ mL}}$. In the denominators we want to convert hours to minutes.

Now starting with 2400 mL/24 hours, set up the conversions so the units cancel:

$$\frac{2400 \text{ mL}}{24 \text{ hr}} \times \frac{15 \text{ drops}}{1 \text{ mL}} \times \frac{1 \text{ hr}}{60 \text{ min}} = \frac{25 \text{ drops}}{1 \text{ min}} \text{ or } 25 \text{ drops/min}$$

Why? Why? XY
Simple Algebra

The algebra in your chemistry class is not difficult. In most cases you simply want to isolate a variable on one side of an equation. A variable is a letter that holds the place for a number. Sometimes you solve the equation to find the numerical value of a variable. Often in chemistry you will "plug" in numbers in place of the variables in formulas.

The letters used for variables can be x or y, just like in an algebra class. But often the letters used in chemistry are taken from what they represent. For example, P stands for pressure, V for volume, T for temperature, etc. Later you will see variables such as $[H^+]$. But no matter how the variable looks, the rules are the same.

Looking For Mr X

Let's look at a few simple equations and see what is going on.
$$x + 9 = 17$$

This equation says when we add 9 to a number, the result is 17. Of course you can see that the number has to be 8. Now let's "solve" the equation by getting the variable alone on one side of the equal sign and everything else on the other side. If we have a number that is added to or subtracted from the variable, to move that number to the other side of the equal sign all we have to do is perform the opposite operation on both sides of the equation. That is, to move a number that is added, subtract it from both sides; to move a number that is subtracted, add it to both sides.

$$x + 9 = 17$$

$$x + 9 - 9 = 17 - 9 \qquad \text{subtracting 9 from both sides}$$

$$x + 0 = 8$$

$$x = 8$$

We can check our answer by plugging it back into the original equation. If we get a true statement, our solution is correct.

$$8 + 9 = 17$$

$$17 = 17$$

What if we make a mistake and get 10 as an answer? Plug in 10 for x:

$$10 + 10 = 17$$

$$20 \neq 17$$

The equation does not check. Therefore this answer is incorrect.

Now let's try another equation.

$$3x = 24$$

This equation states that 3 times some number x is 24, or that some number tripled is 24. A number written in front of a variable is called a coefficient and means that we are multiplying the variable by it. Again in this simple case we can see that x is holding the place for the number 8. Now let's solve. Since $3x$ means 3 times x, we must do the opposite operation, division, to move the 3.

$$3x = 24$$

$$\frac{3x}{3} = \frac{24}{3} \qquad \text{Dividing both sides by 3}$$

$$1x = 8$$

or simply

$$x = 8$$

because the coefficient "1" is understood to be there.

If we plug 8 back into the equation we get a true statement, therefore the answer is correct.

$$3(8) = 24$$

$$24 = 24$$

Now let's look at an equation that includes both multiplication and addition.

$$3x + 5 = 23$$

$$3x + 5 - 5 = 23 - 5 \qquad \text{subtracting 5 from both sides}$$

$$3x = 18$$

$$\frac{3x}{3} = \frac{18}{3} \qquad \text{dividing both sides by 3}$$

$$x = 6$$

Again, we can check our answer by plugging in 6 for the x in the original equation.

$$3(6) + 5 = 23$$

$$18 + 5 = 23$$

$$23 = 23$$

Since we end up with a true statement, the solution is correct.

Now let's try this process with an equation you will see later that converts between Fahrenheit and Celsius degrees.

$$F = 1.8C + 32$$

This equations has two variables, F and C and, as it is, is solved for F. It tells us to get F, we multiply C by 1.8 and then add 32. Now let's solve this equation for C.

$$F = 1.8C + 32$$

$$F - 32 = 1.8C + 32 - 32 \qquad \text{subtracting 32 from both sides}$$

$$F - 32 = 1.8C$$

Next we divide both sides by 1.8. When we divide the left side by 1.8, we divide the entire expression $(F - 32)$.

$$\frac{F - 32}{1.8} = \frac{\cancel{1.8}C}{\cancel{1.8}} \qquad \text{dividing by 1.8}$$

$$\frac{F - 32}{1.8} = C$$

You can always switch an equation around if you like having the variable you have solved for on the left side.

$$C = \frac{F - 32}{1.8}$$

That's about as hard as the solving of equations is going to get in your class.

Half the Equations You Will See
Direct and Inverse Proportions

Many times during the term you will discuss relationships between variables. The two principle relationships are direct and inverse.

If increasing one number causes another number to also increase, you have a direct proportion or variation. Examples in everyday life of direct relationships are:

- The more hours you work, the more money you'll earn.

- The fewer hours you study, the poorer your chemistry grade is likely to be.

- The more positive your attitude, the more likely you will succeed in chemistry.

A direct relationship is like two people roped together mountain climbing. When one person goes up, the other person also goes up.

If one number increasing causes another number to decrease, you have an inverse proportion or variation. Some examples of inverse relationships from life are:

- The faster you drive, the less time it takes to get somewhere.

- The more you drink, the less coherent you will be.

- The more hours you work, the less time you will have to study.

You can also state inverse relationships in the opposite way.

- The slower you drive, the more time it takes to get somewhere.

- The less you drink, the more coherent you will be.

- The fewer hours you work, the more time you will have to study.

An inverse proportion is like two people on a seesaw. When one goes up the other must go down. (Figure 7.1)

On the Up and Up
Direct Proportions

If two variables are directly proportional we can write

$$x \propto y$$

which you read as

$$x \text{ is "proportional to" } y.$$

To turn this statement into an equality, we write

$$x = \text{constant} \times y$$

which says x is equal to a constant times y. The constant may be a number that we choose, a number that we know, a number that is given to us, or a number we have to figure out.

Instead of the word "constant," we use the letter "k" in the equation to make it a little easier to write.

$$x = ky$$

We put the constant in the equation because one number can go up at a different rate than the other.

Let's go back to pay. If you make $1 an hour, your pay will go up exactly at the same rate as the number of hours you work. If you work 20 hours, you will earn $20; if you work 40 hours, you will earn $40, etc. But hopefully you make more than $1 an hour. Suppose you make $8 an hour. If you work 20 hours at $8 an hour, you will earn $160; if you work 40 hours, you will earn $320, etc. In this case our constant, k, is 8. So to calculate your pay for any given number of hours worked at $8 per hour, you can rewrite the general equation replacing k with 8.

$$x = ky$$

$$x = 8y$$

Now you can use the equation to calculate your pay for 20 hours.

$$x = 8(20)$$

$$x = 160$$

You can use the same equation to calculate your pay for 40 hours.

$$x = 8(40)$$

$$x = 320$$

As the number of hours increases, pay increases in this direct relationship.

Now let's take our equation for direct variation and solve for k by dividing both sides by y.

$$x = ky$$

$$\frac{x}{y} = \frac{k\cancel{y}}{\cancel{y}} \qquad \text{dividing both sides by y}$$

$$\frac{x}{y} = k$$

Now suppose we are given that k is 0.5. We would then have

$$\frac{x}{y} = 0.5$$

Look at the relationship between x and y. Can you see that if $x = 1$, y must equal 2 for the answer to be 0.5?

$$\frac{1}{2} = 0.5$$

Notice what happens to y when x is increased to 3 and then to 6:

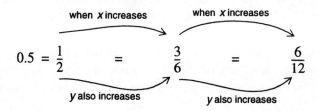

When one variable increases, the other variable also must increase.

In chemistry we use subscripts, small numbers in the lower right hand corner of a variable, when we want to show that a value of a chemical property such as temperature or pressure is changing. For instance, T_1 stands for the first (initial) temperature and T_2 stands for the second (final) temperature.

We just saw that both $\frac{1}{2}$ and $\frac{3}{6}$ are equal to 0.5. So we can say

$$\frac{1}{2} = \frac{3}{6}$$

and if we compare

$$\frac{1}{2} = \frac{3}{6} \quad \text{to} \quad \frac{x_1}{y_1} = \frac{x_2}{y_2}$$

we can see that x_1 is holding the place for 1, x_2 is holding the place for 3, y_1 is holding the place for 2, y_2 is holding the place for 6.

Solving Formulas

In the equation we just derived,

$$\frac{x_1}{y_1} = \frac{x_2}{y_2}$$

which is the general formula for a direct relationship, we have four variables, x_1, x_2, y_1, and y_2. If we are given any three of these variables, we can solve for the fourth. Suppose in a problem you are given that x_1 is 2, y_1 is 5, and y_2 is 15. What is x_2?

We want to solve for x_2, which means we want isolate it on one side of the equation. Therefore we must move y_2 to the opposite side of the equation by doing the opposite operation. Since y_2 is in the denominator, it is being divided, so we must multiply to move it to the other side of the equal sign.

$$\frac{x_1}{y_1} = \frac{x_2}{y_2}$$

$$y_2 \times \frac{x_1}{y_1} = \frac{x_2}{\cancel{y_2}} \times \cancel{y_2} \qquad \text{multiplying both sides by } y_2$$

$$y_2 \times \frac{x_1}{y_1} = x_2$$

We can switch the sides of the equation so that the solved variable is on the left:

$$x_2 = y_2 \times \frac{x_1}{y_1} \qquad \text{which we can also write as} \qquad x_2 = \frac{y_2 \times x_1}{y_1}$$

Plugging in 2 for x_1, 5 for y_1 and 15 for y_2, we get

$$x_2 = 15 \times \frac{2}{5} = 6 \qquad \text{or} \qquad x_2 = \frac{15 \times 2}{5} = 6$$

Notice it doesn't matter whether you write the right side of the equation as two separate fractions or as one fraction.

When x tripled, in this direct relationship, from 5 to 15, y also tripled from 2 to 6.

Cross Multiplying—A Nice Shortcut

Notice in the previous example that when we multiplied both sides by y_2, it moved from the denominator of the right side to the numerator of the left side. We can generalize this procedure into a simple shortcut called "cross multiplying":

We can move a number or variable from the denominator on one side of an equation to the numerator of the other side. Likewise, we can move a number or variable from the numerator of one side of an equation to the denominator of the other side.

In the relationship

$$\frac{x_1}{y_1} \diagup\!\!\!\!\diagdown \frac{x_2}{y_2}$$

we can cross multiply both y_1 and y_2 to get

$$y_2 \times x_1 = x_2 \times y_1$$

Cross multiplying both variables can make it easier to solve problems such as the next example where the variable we are solving for is in the denominator.

Suppose you are given in a problem that x_1 is 7, x_2 is 28 and y_1 is 3. What is the value of y_2?

This question could also be phrased: if x changes from 7 to 28, and y is initially 3, what is the new value of y?

When you work problems like this, it is helpful to make a list of the variables:

$$x_1 = 7 \qquad x_2 = 28$$
$$y_1 = 3 \qquad y_2 = \,?$$

We can observe a lot by looking at the variables before we put them into the equation. Notice that the x values are quadrupling, so the y value will also quadruple.

Solve the equation for y_2:

$$\frac{x_1}{y_1} = \frac{x_2}{y_2}$$

$$y_2 \times x_1 = x_2 \times y_1 \qquad \text{cross multiplying } y_1 \text{ and } y_2$$

$$\frac{y_2 \times x_1}{x_1} = \frac{x_2 \times y_1}{x_1} \qquad \text{dividing by } x_1$$

$$y_2 = \frac{x_2 \times y_1}{x_1}$$

Plugging in the values from the list for the variables we get:

$$y_2 = \frac{28 \times 3}{7} = 12$$

Again, just as we noted, as x quadrupled from 7 to 28, y also quadrupled from 3 to 12.

Looking Ahead

Later in the semester, in the chapter on gas laws, you will study "Charles' Law." Charles' Law states that the volume, or the amount of space, that a gas takes up is directly proportional to the temperature of that gas. That is, as the temperature increases, the volume of the gas increases; as the temperature decreases, the volume of the gas decreases.

If you have ever bought a helium balloon and carried it out of the store on a very cold winter day, you may have noticed that as the helium inside the balloon cools, its volume decreases and the balloon starts to collapse. But don't panic. When you take the balloon back inside where it is warmer, the volume of the helium will increase and the balloon will expand again.

Since Charles' law is a direct variation or proportion, we can write the relationship using our general equation for direct variation, replacing the x and y with V and T:

$$x = ky \qquad \text{becomes} \qquad V = kT$$

V stands for volume, T for temperature and k is a constant. Again this equation simply states that as temperature increases, the volume of gas increases.

We can then solve for k just like we did before in the general equation.

$$V = kT$$

$$\frac{V}{T} = \frac{k\cancel{T}}{\cancel{T}} \qquad \text{dividing both sides by T}$$

$$\frac{V}{T} = k$$

Just like we did for x and y we can write:

$$\frac{V_1}{T_1} = \frac{V_2}{T_2}$$

V_1 and T_1 are initial (or beginning) volume and temperature

V_2 and T_2 are the final volume and temperature.

Let's say the initial volume (V_1) is 2 liters and the initial temperature is 300K. If the temperature (T) doubles to 600K, how will the volume (V) change?

We are given three values, V_1, T_1, and T_2. We have to find the fourth.

First we list our variables to help keep track of them:

V_1 = 2 liter $\qquad V_2$ = ?

T_1 = 300K $\qquad T_2$ = 600K

We solve for the fourth variable, V_2:

$$\frac{V_1}{T_1} = \frac{V_2}{T_2}$$

$$T_2 \times \frac{V_1}{T_1} = V_2 \qquad \text{or} \qquad \frac{T_2 \times V_1}{T_1} = V_2$$

Now plug in the values for the variables

$$V_2 = 600 \times \frac{2}{300} \qquad \text{or} \qquad \frac{600 \times 2}{300} = 4 \text{ L}$$

As the temperature doubled, so did the volume.

Equations or formulas are just simple ways of showing relationships that otherwise take many words to explain.

Ride My Seesaw
Inverse Proportions

In inverse proportions, one variable increasing causes another variable to decrease and vice versa, just like two people on a seesaw (Figure 7.1). We can state an inverse proportion or variation as

$$x \propto \frac{1}{y}$$

This simply says that as y increases, x must decrease. Remember that as the denominator of a fraction increases, the size of the fraction decreases. Here are some examples:

$$\frac{1}{2} = 0.5 \qquad \frac{1}{4} = 0.25 \qquad \frac{1}{8} = 0.125$$

Figure 7.1

As you slice a pie into more pieces, each piece gets smaller. Just as we did for direct proportions, we put in a constant because the numerator will not always be 1. So we then have

$$x = \frac{k}{y}$$

Let's just pick a number for k. Suppose k is 2. The relationship then becomes

$$x = \frac{2}{y}$$

If we choose different values for y, we can see how x changes.

y	1	2	3	4	5
x	2	1	0.667	0.5	0.4

You can see that as y increases, x decreases.

Now let's solve for k.

$$x = \frac{k}{y}$$

$$xy = \frac{ky}{y} \qquad \text{dividing both sides by y}$$

$$xy = k$$

Suppose k is 24. We would then have

$$xy = 24$$

Now if x is 2, what is y?

$$2y = 24$$

$$\frac{2y}{2} = \frac{24}{2} \qquad \text{dividing both sides by 2}$$

$$y = 12$$

Now suppose we increase x to 4. What happens to y?

$$4y = 24$$

$$\frac{4y}{4} = \frac{24}{4} \qquad \text{dividing both sides by 4}$$

$$y = 6$$

When x doubled from 2 to 4, y decreased by half from 12 to 6. Always in an inverse relationship, when one value increases, the other value always decreases.

Solving Formulas Again

We just saw that both 2×12 and 4×6 are equal to 24. So we can say

$$2 \times 12 = 4 \times 6$$

We can replace 2 and 12, x_1 and y_1, and replace 4 and 6, x_2 and y_2. So we can write in general:

$$x_1 y_1 = x_2 y_2$$

This is the general formula for an inverse proportion or variation. Again in problems you will see later on, you will be given three of the variables and asked to find the fourth.

Suppose that in a problem you are given x_1 is 2, x_2 is 10 and y_1 is 20. What is the value of y_2?

This problem could also be stated another way: If y initially is 20 and x changes from 2 to 10, how does y change?

Before we solve the equation, we make a list of variables:

$$x_1 = 2 \qquad\qquad x_2 = 10$$

$$y_1 = 20 \qquad\qquad y_2 = ?$$

Now solve for y_2.

$$x_1 y_1 = x_2 y_2$$

$$\frac{x_1 y_1}{x_2} = \frac{x_2 y_2}{x_2} \qquad\qquad \text{dividing both sides by } x_2$$

$$\frac{x_1 y_1}{x_2} = y_2$$

We put in the values for the variables:

$$y_2 = \frac{2 \times 20}{10} = 4$$

When x increased by a factor of 5 from 2 to 10, y decreased by a factor of 5 from 20 to 4.

Let's try another example.

If y is initially 15 and x decreases from 44 to 17, what is the new value for y?

Again, before we solve, we'll make a list of the variables:

$$x_1 = 44 \qquad\qquad x_2 = 17$$

$$y_1 = 15 \qquad\qquad y_2 = ?$$

Now we solve for y_2.

$$x_1 y_1 = x_2 y_2$$

$$\frac{x_1 y_1}{x_2} = y_2 \qquad\qquad \text{cross multiplying } x_2$$

Then we plug in the values for the variables:

$$y_2 = \frac{(44)(15)}{17} = 38.8235$$

We round the answer to 39 (2 sig figs). Notice that as x decreased from 44 to 17, and y increased from 15 to 39.

Looking Ahead

Later in the class in the chapter on gases you will study "Boyle's Law" which states that as the volume of a gas gets smaller, the pressure of the gas increases. You can feel this if you push down on a bicycle pump with your finger plugging the nozzle. As you compress the air in the piston into a smaller space, you can feel the pressure increase as it gets harder and harder to push down. (See how much chemistry you already know!)

We can state this inverse relationship as

$$P_1 V_1 = P_2 V_2$$

Notice we have just replaced the $x's$ and $y's$ in the general formula for inverse proportions with $P's$ and $V's$.

Now suppose that you are told the pressure of a gas in a piston is 1 atmosphere (about the pressure you feel at sea level) and the volume of the gas is 10 liters. If the volume is decreased to 1 liter, what happens to the pressure?

This problem says the initial pressure P_1 is 1 atm ("atm" is the abbreviation for atmosphere) and the volume decreases from an initial value V_1 of 10 liters to a final value V_2 of 1 liter. What is the resulting pressure P_2?

We make a list of the values to help keep them straight:

$$P_1 = 1 \text{ atm} \qquad P_2 = 10 \text{ L}$$
$$V_1 = 10 \text{ L} \qquad V_2 = 1\text{L}$$

Now solve for P_2:

$$P_1 V_1 = P_2 V_2$$

$$P_2 = \frac{P_1 V_1}{V_2}$$

cross multiplying V_2

Plug in the values for the variables:

$$P_2 = \frac{(1\,\text{atm})(10\,\text{L})}{1\,\text{L}} = 10\,\text{atm}$$

As the volume decreased by 10, the pressure increased by 10.

All the equations in the chapter on gases will be variations of direct and inverse proportions.

Looking Ahead

Later in the class in the chapter on solutions, you will see the "dilution equation."

$$M_1 V_1 = M_2 V_2$$

This is another example of an inverse proportion. M stands for "molarity," V stands for the volume of solution. You do not need to know what molarity is to see the relationship. It has the same form as Boyle's Law. As the molarity M increases, the volume V decreases and vice versa.

Looking Ahead

In the chapter on acids and bases you will see one more inverse relationship. It looks like this:

$$[H^+][OH^-] = 1 \times 10^{-14}$$

$[H^+]$ and $[OH^-]$ are variables, just like x and y. If we compare to the general formula,

$$xy = k$$

you can see $[H^+]$ is equivalent to x, $[OH^-]$ is equivalent to y and the constant k is 1×10^{-14}. $[H^+]$ represents the concentration of acid, $[OH^-]$ stands for the concentration of base. You don't need to know what acids and bases are right now. This relationship simply says that as something gets more acidic, it becomes less basic and vice versa. It does not matter if the variables are a little strange looking. Let's put some numbers in and see what happens.

In pure water, $[H^+]$ is 1×10^{-7}. What is $[OH^-]$?

We solve for $[OH^-]$:

$$[H^+][OH^-] = 1 \times 10^{-14}$$

$$[OH^-] = \frac{1 \times 10^{-14}}{[H^+]} \qquad \text{cross multiplying } [H^+]$$

Putting in 1×10^{-7} for $[H^+]$:

$$[OH^-] = \frac{1 \times 10^{-14}}{1 \times 10^{-7}} = 1 \times 10^{-7}$$

This is the concentration where the acid equals the base. Pure water is neither acidic nor basic.

Now what if we squeeze some lemon juice into our water and $[H^+]$ increases to 1×10^{-5}. What happens to $[OH^-]$?

$$[OH^-] = \frac{1 \times 10^{-14}}{1 \times 10^{-5}} = 1 \times 10^{-9}$$

When $[H^+]$ increased, $[OH^-]$ went down. Remember, the larger the negative exponent, the smaller the number.

Notice that in proportion equations, there is no adding or subtracting, just multiplying and dividing.

That's It For Algebra

Well, that's it for algebra. It doesn't get more complicated than this. As you've seen in this section, formulas are not just algebra and numbers but relationships you can visualize and understand.

Now try the practice problems on page 173.

Learning to Relax

It's hard to study effectively and you certainly won't perform your best on an exam if your body is overly tense. Also when you are tense, things always look worse than they are. However, some stress is normal when you enter into the unknown. And a little tension keeps you from getting too complacent and making careless errors. It is the excess tension that you need to be able to control.

There are three factors you can consciously control to reduce your stress and increase the effectiveness of your studying: posture, muscle tension and breathing. We will discuss just a few techniques you can try with each factor. There are many books available on relaxation; we encourage you to seek out these resources.

Posture

When we are concentrating hard, especially when we are tired, we often tend to pull our neck down into our shoulder blades, collapse our chest and let the lower spine bow out. Sitting with the body in this position tends to make us tired and tense. Our muscles work overtime to hold the body up and the lungs cannot breathe efficiently.

When you are studying, and especially when you are taking a test, imagine a string is tied to the crown of your head lifting it gently away from your shoulder blades and elongating your neck and spine. When you are sitting or standing, imagine your spine gently pulled upward. Think tall but not rigid. When you sit at your full height, it will help you will feel at ease and confident.

Muscle Tension

A tense muscle uses more energy and requires more oxygen and nutrients than a muscle at rest. But a tense muscle constricts the capillaries that deliver oxygen and nutrients to it, which leads to fatigue of the muscle. The exhausted muscle becomes even more tense starting a downward cycle. There are three things you can do to relieve muscle tension: stretching, massage and movement.

While you are studying, remember to stretch. Stretch your hands above your head. Clasp your hands behind your back and expand your chest. Stand up periodically and stretch any way that feels good. If you are at home, lie on the floor and stretch your arms over your head.

Gently massage your face and scalp. Then massage your neck and shoulders. Massaging your feet will refresh you during a long study session.

Take a walk. Walking is one of the best ways to relax. Even a short walk around the block will help you relax and feel more alert.

Periodically look out the window or across the room to relax your eye muscles. You can also rub your hands together to create a gentle friction and then place your the hollows of your palms over your eyes. The warmth will relax your eye muscles.

Breathing

Your breath is tied to your emotions. Have you noticed that after a traffic skirmish your breath becomes rapid and shallow. This is one symptom of what is known as "fight or flight," a remnant of the days when we had to either confront a danger or run. In the modern world this reaction does little to help us. The next time a traffic incident occurs, take a long slow breath and feel what happens. You should feel more calm and in control.

The same "fight or flight" reaction can happen on a smaller scale when you are studying unfamiliar material or when you are taking an exam. Here are some breathing techniques you can try anytime you feel tense.

Breathe in slowly through your nose as you count to five. Then breathe out slowly. Pause for a second when you have completely exhaled, then repeat for a few more breaths. When you inhale, expand your abdomen first to fill the lower lungs. Then let your chest rise to fill the upper lungs. Don't force the breath; let it come easy and natural. You can do this exercise a few minutes before you take an exam.

You can write "breathe" on the top of your exam to remind you to take breathe deeply while taking a test.

CHAPTER 8

Light Headed With All These Heavy Thoughts—Density

"This stone is as heavy as lead," Diane said to Tim. "Yeah, but pick up this balsa wood. It's almost as light as styrofoam," Tim replied.

Have you made comments like these? Or have you heard a question like: "Which is heavier, a pound of lead or a pound of feathers?" Of course this is a trick question; they both weigh one pound! But the question does make you think! Often, when we use the words "heavy" and "light" in everyday life, we are really talking about what is referred to in chemistry as density. Density is the amount of mass per volume of a substance.

You will be happy to know that there are only three types of number problems involving density. Once you are familiar with them, they are easy to recognize and solve. But before we look at density in number problems, let's see if we can develop your intuitive understanding of what density is. Again, just like with metric conversions, if you can visualize density so that it is not just abstract mathematics, you will not only be able to recognize and solve density problems, but you will also have a powerful concept to help you think about things that occur in everyday life.

If we translate the definition of density as mass per volume into a formula, we get:

$$\text{Density} = \frac{\text{mass}}{\text{volume}}$$

As you saw in Chapter 2, the "per" in the definition is a fraction bar that means to divide, just as miles per gallon means miles divided by gallons (miles/gallons). Mass is in the numerator, and volume is in the denominator. In chemistry the units used most often in density problems are grams per milliliter, $\frac{g}{mL}$ or grams per cubic centimeter, $\frac{g}{cm^3}$.

Remember that milliliters and cubic centimeters are identical ($1 \text{ mL} = 1 \text{ cm}^3$).

Reading the Tables

Every pure substance has its own particular density. Table 8.1 lists the densities of some common substances. You can see that densities vary. The density of water is $\frac{1.00g}{mL}$ and the density of iron is $\frac{7.87g}{mL}$. One milliliter of water weighs 1.0 gram while one milliliter of iron weighs a much heavier 7.87 grams. Air, on the other hand, is very "light." One milliliter of air weighs only 0.001184 grams.

It is no coincidence that water has a very convenient density of $\frac{1.00g}{cm^3}$. When the French invented the metric system, they defined a gram to be the mass of water that would fit in a one-centimeter cube (1 gram of water $= 1 \text{ cm}^3$).

Look at the units in Table 8.1. The units listed at the top of the table are g/mL or g/cm^3. The table in your textbook will list the units either way. Remember a mL is the same as a cm^3. Also notice the units in tables are listed with a slanted denominator line. Let's look up the density of alcohol in Table 8.1. The number listed for alcohol is 0.785 and the units at the top of the table are g/mL or g/cm^3. If we want to use this density to solve a problem, we will write $\frac{0.785g}{1 \text{ mL}}$ or $\frac{0.785g}{1 \text{ cm}^3}$. Remember, in mathematics, "1's" by themselves in the denominator are almost never shown. They are left implicit. But to avoid confusion, we add the "1" to the volume. The density is 0.785 grams per *one* milliliter.

Table 8.1: Densities of Some Common Substances

Substance at 25°C	Density g/mLor g/cm^3	Substance at 25°C	Density g/mLor g/cm^3	Substance at 25°C	Density g/mLor g/cm^3
Balsa wood	0.12	Cork	0.26	Aluminum	2.70
Pine	0.4	Olive oil	0.92	Iron	7.87
Maple	0.75	Gasoline	0.66	Silver	10.5
Human fat	0.94	Ethyl alcohol	0.785	Lead	11.3
Bone	1.8	Ice (at 4°C)	0.92	Mercury	13.6
Muscle	1.08	Water	1.00	Gold	19.6
Urine	1.003-1.030	Helium	0.000179	Carbon dioxide	0.00196
Plasma (blood)	1.03	Air	0.001185	Hydrogen	0.00090

Hot Air Rises
Density and Temperature

The densities listed in tables have to be reported at a particular temperature because density varies inversely with temperature. As temperature increases, density decreases. When any substance is heated, it expands in volume. Because volume is in the denominator in the definition of density, as the volume increases the density decreases. (The fraction 3/8 is smaller than the fraction 3/4.) For solids and liquids the changes are small. The density of iron decreases from 7.87 g/cm^3 at 25°C to 7.544 g/cm^3 at 100°C and the density of water decreases from 1.0 g/mL at 0°C to 0.92 g/mL at 100°C.

Gases change in density more dramatically. The density of air decreases from 0.001185 g/mL at 25°C to 0.0009456 g/mL at 100°C, a decrease of 20% compared to the 4.2% decrease in the density of liquid water over the same temperature range.

Cream Rises

Less dense substances float on more dense substances. Because it is less dense, oil floats on water. Likewise wood floats on water, iron sinks. Substances with densities less than 1.0 g/mL float on water while substances with densities greater than 1.0 g/mL sink. Ice is less dense than water because, unlike almost every other substance, water expands when it freezes into ice, that is, the volume gets bigger in the density definition. This expansion of ice is one of the principal causes of cracks in concrete and chuck holes in roads. We are fortunate that ice is less dense than water. That floating ice forms an insulating cover that keeps the water underneath from freezing. The ice protects the fish and the other creatures that live in the water underneath.

Muscle has a density of 1.08 g/mL, while fat has a density of 0.92 g/mL. The more fat someone has, the better he or she will float. More muscular people are better at sinking. But if the body builders among us really want to float, they can go to the ocean. Because of the salt content, ocean water has a density of 1.12 g/mL. So even they can float a little better there.

The less dense air in the hot air balloon floats on the colder, more dense air outside the balloon, just like the less dense helium in a helium balloon tries to float above the air, only the string holds it down. The differences in densities between hot and cold air cause hot air to rise and cold air to sink. This movement of air masses creates wind and makes rain, storms and other weather patterns.

Lead and Feathers Again

Imagine that you have a two-pound piece of lead in your left hand and two pounds of feathers (a feather pillow) in your right hand. Picture in your mind how each looks. Can you visualize the lead fitting into the palm of your hand and the feathers as a big fluffy pillow? If one material is denser than another, an equal mass of the denser material (the lead, in this case) will fit into a smaller space. A material that is more dense will be compact and "heavy." A material that is less dense will be bulky and "light." Remember that "heavy" and "light," as used in everyday language often refer to differences in density rather than to actual weights.

Now again imagine a one-pound lead bar in your left hand, but this time in your right hand imagine a one-pound gold bar. (Imagining a gold bar in your hand might make for a pleasant daydream.) Which bar is larger? That is, which bar has a greater volume? Look up the densities in Table 8.1 and think back to the feathers and the lead. The less dense feathers occupied more space. The lead is less dense than the gold, so the one pound lead bar will be larger than the one pound gold bar. Or, we can say that the more dense gold will be more compact than the lead. So when we say that something is as heavy as lead, we could really say that it is as heavy as gold. But then how many of us have ever held a chunk of gold in our hands?

One way of measuring your percent of body fat is to completely submerge you in water and measure how much the water rises. We measure your body's volume by displacement. Since fat is less dense than muscle it will occupy more space. If two people with the same mass are submerged, the person with more fat will displace more water.

When starting an exercise program, people are often dismayed to find that initially, while they lose inches around the waist (volume that is), they gain weight because more dense muscle is replacing less dense fat!

Can You Speak Density in English?

We can think of density in everyday life in the English units of pounds per gallon. A gallon of water weighs 8.2 pounds, so we say that the density of water is 8.2 lb/gal. If you carry two gallons of milk (which is mostly water), you are hauling a little over sixteen pounds. If you go backpacking or on a long hike, every quart (1/4 gallon) of water you carry will weigh about two pounds. Water has the same density even when it is a part of food, which is why dehydrated food is popular among backpackers—all that "heavy" water is removed.

Equal Volumes Of Different Densities

What if you have equal volumes of two different substances? If you have a liter of water and a liter of the liquid metal mercury, which is heavier? Compare their densities in Table 8.1. The liter of mercury is 13.6 times heavier than the liter of water. If you have equal volumes, the denser substance will weigh more.

Imagine that your friend is carrying a fifth of 190 proof vodka (which is mostly ethyl alcohol) and you have a "fifth" of water. Which of you has the greater burden? Find the densities of water and ethyl alcohol in Table 8.1. It looks like you have the heavier load. Water is denser than ethyl alcohol (1.0 g/mL vs. 0.79 g/mL). It does not matter what the actual volume of a "fifth" is (it is 1/5 of a gallon or about 750 mL). The important point is that as long as the volumes are equal, the denser substance will weigh more.

Likewise, firewood is sold in a volume unit called a "cord." A cord is a volume of wood that is four feet high, four feet wide, and eight feet long, or 128 cubic feet. Sugar maple, which is a dense wood has almost twice the heating capacity of basswood, which is a "light," not very dense wood. If you bought cords of sugar maple and basswood, the cord of sugar maple world weigh more and supply more heat than the cord of basswood.

Supermarket Density

Let's look at some examples of density in the supermarket. Many products you buy are sold by weight. You buy cheese, bread, and cereal by the pound. Often labels on products such as crackers, potato chips or breakfast cereal have a statement such as "sold by weight, not by volume." Products like chips or corn flakes settle during storage and occupy a smaller volume, that is, they become more dense. The manufacturers want to remind you that they are selling you 12 ounces even though the box does not appear to be full.

Other products such as milk or ice cream are sold by volume, that is pints, quarts or half gallons. You buy a half gallon of milk or a quart of ice cream. Ice cream is interesting because, even though it is a solid (at least while it is cold), it is sold by volume. Being good businessmen, the ice cream manufacturers want to sell you the smallest amount (mass) of ice cream possible in that quart. Next time you are at the supermarket take a few half gallons of different brands of ice cream to the produce department and weigh them. You will see some differences in the pounds per half gallon of different brands. The easiest thing the ice cream manufacturer can do to decrease the density of the ice cream is to add an ingredient that is very "light." The manufacturer has a large, inexpensive supply of that substance, air. They whip air into the ice cream before it is frozen. It does not take much; remember air has a density of 0.001185 g/mL. So the next time you buy ice cream, weigh that bargain half gallon of ice cream and compare to the more expensive brand. You may not be getting much of a bargain at all!

Life and Death Density
The Story of Archimedes

Before we look at number problems, let's look at an old story that involves density. Archimedes, a Greek mathematician who lived in Syracuse around 300 BC, was somewhat of a court scientist for the King of Syracuse. The king decided he wanted to have a gold crown made. He measured out an amount of gold with which to make the wreath and gave it to the court goldsmith. Gold was as expensive then as it is now, and the king watched every ounce of it. When the goldsmith completed his work, the king suspected that he had stolen some of the gold and had substituted a cheaper and less dense metal like copper or silver, but he could not prove it. So he gave the problem to Archimedes.

Archimedes thought for a long time. One morning as he got into his bathtub, Archimedes noticed something we all have seen. The water in the tub rose. He jumped up and ran naked through the streets of Syracuse shouting, "Eureka. Eureka," which means, "I've found it. I've found it." He had found the solution.

When he lowered himself into the water, his body displaced the water and the level rose. So he weighed the crown. Then he weighed out an equal mass of pure gold. He put the gold in a jar of water and measured how high the water rose. He immersed the crown in a jar of the same size and showed the king that the water rose higher. The crown had a greater volume than an equivalent mass of pure gold. Therefore the crown had to contain some other metal less dense than gold. (Remember how less dense feathers takes up more space than lead or how a person with more fat displaces more water than a muscular person.) The goldsmith lost his head. Would you have thought density could be a matter of life and death!

Luckily, any troubles you have with chemistry are just a little smaller than the unfortunate goldsmith's. Now that you have some intuitive understanding of density, let's look at the three types of density problems.

First Type of Density Problem
Given Mass and Volume, Find the Density

In the first type of density problem, you are given a mass and a volume and asked to find the density. To solve this type of problem, you use the definition of density and divide the mass by the volume.

$$\text{Density} = \frac{\text{mass}}{\text{volume}} \quad \text{with units of} \quad \frac{g}{mL} \quad \text{or} \quad \frac{g}{cm^3}$$

Suppose you measure out 10.0 mL of hexane and you measure its mass to be 6.831 grams. What is the density?

Divide the mass by the volume.

$$\text{Density} = \frac{\text{mass}}{\text{volume}} = \frac{6.831 \text{ g}}{10.0 \text{ mL}} = \frac{0.6831 \text{ g}}{\text{mL}} \text{ or } 0.683 \text{g/mL}$$

Notice that we round our answer to three significant figures.

Suppose that this is a problem from a test:

A metal block has a volume of 5.52 cm^3 and weighs 44.432 grams. What is the density?

Again density is mass divided by volume:

$$\text{Density} = \frac{\text{mass}}{\text{volume}} = \frac{44.432 \text{ g}}{5.52 \text{ cm}^3} = \frac{8.05 \text{ g}}{\text{cm}^3} = 8.05 \text{ g/cm}^3$$

Let's try a problem like the one Archimedes had. We measure the volume of a gold ring to be 0.52 cm^3 by measuring how much water it displaces. We measure the mass to be 7.314 grams. What is the density? Is the ring pure gold?

Using the definition of density

$$\text{Density} = \frac{\text{mass}}{\text{volume}} = \frac{7.314 \text{ g}}{0.52 \text{ cm}^3} = \frac{14 \text{ g}}{\text{cm}^3}$$

Since the density is less than that of pure gold (19.3 g/cm^3), the ring is not pure gold. Most jewelry is not made out of pure gold because pure gold is too soft.

Notice our answer has two significant figures because the volume was only measured to 2 significant figures.

Density as a Unit Conversion

The second and third types of density problems use density as a unit conversion. The density of gold is $\frac{19.3 \text{ g}}{1 \text{ cm}^3}$. Just like any other unit conversion, you can also write $\frac{1 \text{ cm}^3}{19.3 \text{ g}}$.

Second Type of Density Problem: Given Density and Volume, Find the Mass

In a laboratory, often it is easier to measure out a volume of a liquid than it is to weigh it. If you measure the volume, you can use density to find the mass without having to weigh directly on a balance.

In the second type of density problem, you are given a volume and the density and asked to calculate the mass.

What is the mass of 40.0 mL of carbon tetrachloride? The density of carbon tetrachloride is 1.594 g/mL.

The 40.0 mL is our given quantity. We will use density as a unit conversion and set it up so that the units cancel:

$$40.0 \text{ mL} \times \frac{1.594 \text{ g}}{1 \text{ mL}} = 63.8 \text{ g}$$

In this type of problem you have to be given the density or, if you are not given the density, know where to look it up. Certainly on a test you will be given the density, but in some homework problems in your textbook, you may have to look up the density in the table in the textbook.

What is the mass of $27.3 \, \text{cm}^3$ of ethyl alcohol? Find the density in Table 8.1.

The given quantity is $27.3 \, \text{cm}^3$. We look in the table and find the density of ethyl alcohol to be $0.785 \, \text{g/cm}^3$.

$$27.3 \, \cancel{\text{cm}^3} \times \frac{0.785 \, \text{g}}{1 \, \cancel{\text{cm}^3}} = 21.4 \, \text{g}$$

What is the mass of 10.0 mL of water?

Water is the one compound whose density you might be expected to memorize. After all it *is* such an easy number.

$$10.0 \, \cancel{\text{mL}} \times \frac{1.00 \, \text{g}}{1 \, \cancel{\text{mL}}} = 10.0 \, \text{g}$$

Third Type of Density Problem:
Given the Density and Mass, Find the Volume

If we need to know the volume of an oddly shaped solid object, we could measure it by the displacement of water. But if we know the density and the weight of the object, we can calculate its volume without having to displace water.

In the third kind of density problem, you are given density and mass and asked to calculate the volume.

What is the volume of a 25.0 gram block of lead. The density of lead is $11.6 \, \text{g/cm}^3$.

The given quantity is 25.0 grams. We turn the density upside down so that the mass units cancel.

$$25.0 \, \cancel{\text{g}} \times \frac{1 \, \text{cm}^3}{11.6 \, \cancel{\text{g}}} = 2.16 \, \text{cm}^3$$

What is the volume of 27.2 g of diethyl ether? Find the density in Table 8.1.

The given quantity is 27.2 grams. We find the density of diethyl ether to be 0.714g/mL.

$$27.2 \, \cancel{\text{g}} \times \frac{1 \, \text{mL}}{0.714 \, \cancel{\text{g}}} = 38.1 \, \text{mL}$$

Once again we turned density upside down so that the mass units cancel to find the volume.

Which has a greater volume: a 50.0 gram block of lead or a 50.0 gram block of gold?

We find the densities of lead and gold in Table 8.1 to be 11.6 g/cm^3 and 19.3 g/cm^3. Remember how we imagined holding the lead and gold in our hands previously. We saw that since gold is more dense, it will fit into a smaller space. Now let's do the conversions to calculate the exact values.

$$50.0 \text{ g lead} \times \frac{1 \text{ cm}^3}{11.6 \text{ g}} = 4.31 \text{ cm}^3 \text{ lead}$$

$$50.0 \text{ g gold} \times \frac{1 \text{ cm}^3}{19.3 \text{ g}} = 2.59 \text{ cm}^3 \text{ gold}$$

The less dense lead takes up a greater volume than the more dense gold, just like the 2 lbs. of feathers had a greater volume than the 2 lbs. of lead. Even before you plug the numbers into the calculator, you can see that the volume of the gold will be smaller because you are dividing by a larger number.

Sink or Float—Specific Gravity

Sometimes we use the term specific gravity in place of density. Specific gravity is the ratio of the density of a substance to the density of water, usually at the same temperature. Water is that chemical compound that all of us are familiar with, so it is convenient to compare densities to it. The specific gravity of gold is

$$\frac{\dfrac{19.3 \text{ g}}{1 \text{ mL}}}{\dfrac{1.0 \text{ g}}{1 \text{ mL}}}$$

Remember when we divide one fraction by another we multiply by the recipropcal of the bottom fraction. So we get

$$\frac{19.3 \text{ g}}{1 \text{ mL}} \times \frac{1 \text{ mL}}{1.0 \text{ g}}$$

Because all the units cancel, specific gravity is a unitless number. And since water has a density 1.0 g/mL the magnitude of the number remains the same.

The density of gold is 19.3 g/mL, and the specific gravity is simply 19.3. Here you have to be careful. If someone asks you what is the density of gold, technically speaking, you have to answer 19.3 g/mL (or 19.3 g/cm^3). You can not be lazy and leave off the units.

But if someone asks the specific gravity you can simply and happily answer 19.3. Practically speaking, to go from density to specific gravity just chop off the units. To go from specific gravity to density, reattach the units.

Now let's see how specific gravity might be incorporated into problems.

The mass of a chemical is 8.45 grams and the volume is 5.00 mL. What is the density and what is the specific gravity?

This is the first type of density problem where you are given mass and volume and are asked to calculate density.

$$\text{Density} = \frac{\text{mass}}{\text{volume}} = \frac{8.45 \text{ g}}{5.00 \text{ mL}} = \frac{1.69 \text{ g}}{\text{mL}}$$

To get the specific gravity, we just chop off the units. Specific gravity = 1.69.

The specific gravity of olive oil is 0.918. What is the volume of 125 grams of olive oil?

This is an example of the third type of density problem. We are given the mass and asked to find the volume. If the specific gravity is 0.918, the density is 0.918g/mL. We turn the density upside down so that the units cancel.

$$125 \text{ g} \times \frac{1 \text{ mL}}{0.918 \text{ g}} = 136 \text{ mL}$$

Using an Hydrometer

You can measure the specific gravity of a liquid directly with an instrument called a hydrometer. A hydrometer is simply a hollow glass tube with a weight on one end. The tube has marks on the side that are calibrated to read specific gravity. The tube will sink deeper in liquids that have a low specific gravity and sink less deep in liquids with a higher specific gravity. (Figure 8.1)

Wine makers use hydrometers to measure the progress of fermentation. The specific gravity will decrease from 1.082 in grape juice which is 20 to 25% sugar to 0.92 in the completely fermented wine which is about 12 to 15% ethyl alcohol. The hydrometer sinks deeper as the wine ferments.

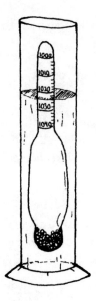

Figure 8.1

Now try the practice problems on page 175.

Preparing for the Test

- Test preparation begins on the first day of the course. Keeping current with all daily assignments is the most effective method of preparing for tests.

- Form a study group. You never really learn anything until you have to explain it to someone else. A study group will force you to do this. Working with others will also help you learn more study strategies. But study on your own both before and after the study group.

- Tests are more difficult than individual assignments because they contain a greater variety of material. To avoid getting confused, you must "overlearn" the material. To "overlearn," study actively until you really feel that you understand the material or can work the problems with no difficulty. Check to see how long you had to study or how many problems you had to work to reach this point. To "overlearn," you must spend at least half as much more time or work half as many more problems. Then you must also spend a small amount of time each day, maybe 10-15 minutes, reviewing past assignments after you have completed the new assignment.

🐿 If there is a concept you do not understand, get help immediately. In chemistry learning is often sequential, i.e., new lessons build on past lessons. Therefore falling behind can be fatal!

🐿 Try to predict test questions as you are studying.

🐿 If practice tests are available, get them as soon as possible and do them at least several days before the test. Then you will still have time to get help if you find there are questions that are giving you difficulty.

🐿 If you do fall behind in your studies, never try to learn new material the day or the night before a test. You will not have time to "overlearn" it. Therefore what will normally happen is you will confuse new material with what you already know. This confusion can cause you to miss even more test questions than you would have missed had you not tried to "cram in" the new material. Concentrate on what you know on the night before the test. Then, after the test is over, catch up immediately.

🐿 The night before the test be sure to get enough rest. Also check your calculator and put it and any other materials you will need in a place where you will be sure to pick them up before leaving for class.

CHAPTER 9

When You're Hot
Temperature Conversions

In America we are accustomed to thinking and feeling in Fahrenheit degrees. We know intuitively that 74°F is a nice day, 95°F is really hot, and –20°F is frigid cold. Most of us know that "normal" body temperature is 98.6°F. But, as you probably know, the rest of the world uses a different temperature scale, the Celsius scale, which is part of the metric system.

This chapter has nothing really to do with chemistry. If the United States used the metric system along with the rest of the world, this chapter would would not be necessary.

In this chapter you will get familiar with the Celsius scale and learn to convert between Celsius and Fahrenheit temperatures. You will also be introduced to the Kelvin scale that you will use later in the class when you study the properties of gases.

What is Temperature?

Temperature is a measurement of how heat flows. (We will talk in the next chapter about what exactly heat is.)

Heat always flows from a warmer body to a cooler body. Let's say one of the "bodies" is your body, and the other "body" is the air; we could even call it the "body of air." When you go outside on a cold winter day, the heat flows out of your body to the cold air rapidly. On a warm summer day, it flows much less rapidly. If it is 85°F outside, there is a less steep gradient between your body's 98.6°F temperature than when it's 20°F outside. You can think of it like a hill. In the winter there is the steep hill running from 98.6°F to 20°F.

In the summer there is the shallow hill of 98.6°F to 85°F. On a hot summer day the "hotness" you feel, unless it's warmer than 98.6°F outside, is not the heat flowing toward you, it is your body heat not being able to flow out of your body.

Comparing Thermometers

It takes two temperatures to define a temperature scale, just as in mathematics it takes two points to determine a straight line. Two convenient points we can use are the boiling point and the freezing point of water. We all know that water boils at 212°F and freezes at 32°F. On the Celsius scale, water boils at 100°C and freezes at 0°C, which is why Celsius is metric: Mr. Celsius chose exactly 100 degrees between the two temperatures. Figure 9.1 shows the relationship between the scales.

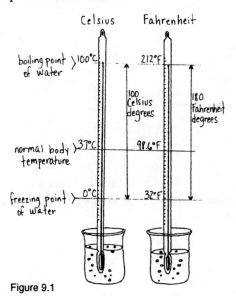

Figure 9.1

Now let's develop a relationship between a Fahrenheit degree and a Celsius degree. To put it another way, we want to write an equation to change a Celsius temperature to a Fahrenheit temperature.

In Figure 9.1 you can see that between the freezing and the boiling points of water there are 100 Celsius degrees (100°C − 0°C) while there are 180 (212°F − 32°F) on the Fahrenheit scale. So we may write

$$180°F = 100°C$$

Dividing both sides by 100°C we get

$$\frac{180°F}{100°C} = \frac{\cancel{100°C}}{\cancel{100°C}}$$

$$\frac{1.8°F}{1°C} = 1$$

This is a conversion that allows us to convert a change in temperature in the Celsius scale to Fahrenheit. For instance if the high temperature on a spring day is 22°C and the low temperature is 7°C. What is this difference in temperature in Fahrenheit degrees?

The difference is 14°C. We will call the Fahrenheit temperature T_F and the Celsius temperature T_C.

$$T_F = \frac{1.8°F}{1°C} \times T_C$$

$$T_F = \frac{1.8°F}{1\cancel{°C}} \times 14\cancel{°C} = 25°F$$

Based on this relationship, we can make a "rule of thumb": there are about twice as many (rounding 1.8 to 2) Fahrenheit degrees as Celsius degrees in a temperature span.

If we want to convert a particular Celsius temperature to Fahrenheit, we have to take into consideration that the freezing point of water is 32 degrees higher on the Fahrenheit scale. So we add 32°F to the relationship.

$$T_F(°F) = \frac{1.8°F}{1°C} \times T_C + 32°F$$

In words this equation says: to convert a Celsius temperature to Fahrenheit, multiply the Celsius temperature by 1.8 and then add 32.

In most chemistry texts this equation is simplified by leaving out T_F and T_C variables and letting the temperature units become the variables. The equation then becomes:

$$°F = 1.8°C + 32$$

We will use this streamlined form to do problems in this chapter. But keep in mind that the previous form of the equation better shows the actual conversion between the temperature units

Now the Bank Thermometer Makes Sense
Converting From C to F

Let's convert some Celsius temperatures to Fahrenheit.

Room temperature is considered 20°C in most chemistry references. What is that temperature in °F?

$$°F = 1.8°C + 32$$

$$°F = 1.8(20) + 32$$

$$°F = 36 + 32 = 68°F$$

In math, we always do multiplication before addition.

On a calculator we press 1.8 ⎡×⎤ 20 ⎡+⎤ 32 ⎡=⎤   The display reads   *68.0*

Twenty years ago, 25°C was considered room temperature. What is the Fahrenheit equivalent?

$$°F = 1.8°C + 32$$

$$°F = 1.8(25) + 32$$

$$°F = 45 + 32 = 77°F$$

Let's try a negative (below zero) temperature. Convert –10°C to °F.

$$°F = 1.8°C + 32$$

$$°F = 1.8(-10) + 32$$

$$°F = -18 + 32 = 14°F$$

The product 1.8(–10) is negative, because (+)×(–) = (–). Then we add –18 and 32. Remember when adding a (–) and (+), subtract the smaller number from the larger and give the answer the sign of the larger number.

To do the problem on a calculator, enter

1.8 ⎡×⎤ 10 ⎡+/-⎤ ⎡+⎤ 32 ⎡=⎤   The display reads   *14.*

Remember to press the $\boxed{+/-}$ button after you enter a number if you want to make the number negative.

What About the Other Direction
Converting From F to C

As we just saw, the equation

$$°F = 1.8°C + 32$$

will let you plug in any Celsius temperature and give you the Fahrenheit equivalent. To go the other way from a Fahrenheit temperature to Celsius, we solve the equation for °C.

$$°F = 1.8°C + 32$$

$$°F - 32 = 1.8°C + 32 - 32$$

$$°F - 32 = 1.8°C$$

$$\frac{°F - 32}{1.8} = °C$$

Or you can switch sides if you like the variable you solved for on the left.

$$°C = \frac{°F - 32}{1.8}$$

This equation tells you to subtract 32 from the Fahrenheit temperature, then divide by 1.8.

Now let's convert some Fahrenheit temperatures to Celsius temperatures.

What is 45°F, the temperature of a moderately cold winter day, in Celsius?

$$°C = \frac{45 - 32}{1.8} = \frac{13}{1.8} = 7.2°C$$

Body temperature is 98.6°F. Convert to °C.

$$°C = \frac{98.6 - 32}{1.8} = \frac{66.6}{1.8} = 37°C$$

On the calculator press: 98.6 $\boxed{-}$ 32 $\boxed{=}$ $\boxed{+}$ 1.8 $\boxed{=}$ The display reads *37.0*

Notice that you pressed $=$ twice. You pressed it the first time to perform the subtraction in the numerator before you divided by 1.8. You can also make the calculator subtract before dividing by using the parentheses buttons on the calculator, that is (98.6 − 32)÷1.8. However it's probably easier to use the $=$ button.

Now let's convert a really cold winter day temperature, −20°F, to Celsius.

$$°C = \frac{-20-32}{1.8} = \frac{-52}{1.8} = -29°C$$

−20 − 32 = −52 (Remember two negatives combine to form a larger negative.) Then

$\frac{-52}{1.8} = -29$, (because a negative divided by a positive is a negative.)

On the calculator

20 +/- · 32 = ÷ 1.8 =   The display reads   - 28.8888888

We round our answer to two significant figures to −29°C.

Now let's try an even colder temperature that you might find up north in Hudson Bay. Convert −40°F to Celsius.

$$°C = \frac{-40+(-32)}{1.8} = \frac{-72}{1.8} = -40°C$$

Did we make a mistake? We came up with the same number we started out with. No, we didn't make a mistake. −40 is the one temperature where both thermometers read the same. Forty below Fahrenheit is 40 below Celsius.

On the calculator

40 +/- · 32 = ÷ 1.8 =   The display reads   -40.

Absolute Temperatures
Converting to Kelvin

There is a third temperature scale that you will use in problems involving gases (where you can't have negative temperatures). It is called the Kelvin scale. The Kelvin scale sets absolute zero as 0. On the Celsius scale absolute zero is −273.15 and is the coldest anything can get. At that temperature all molecular motion stops. The very simple conversion is

$$K = °C + 273$$

(Three significant figures is good enough for most temperature conversions, so we round 273.15 to 273.)

Convert 37°C to Kelvin. It is very easy to do this conversion. Just add 273 to the Celsius temperature.

$$K = 37 + 273 = 310 \text{ K}$$

Note that the ° sign is not used and we simply say 310 Kelvins, not degrees Kelvin. The Kelvin is actually the "official" SI unit for temperature. But most chemists stick to the less official Celsius degrees except in situations where Kelvins must be used. Can you imagine if bank thermometers started giving the Kelvin temperature. It is now 289 degrees outside!

The boiling point of liquid nitrogen (sometimes used to remove skin growths or blemishes) is –196°C. What is it in K?

$$K = -196 + 273 = 77$$

Note Kelvin temperatures are always positive.

To solve for Celsius, simply subtract 273 from both sides of the equation.

$$°C = K - 273$$

Now let's convert 200 K to °C.

$$°C = 200 - 273 = -73°C$$

Temperature is the only unit that you use equations to convert. All other units will be converted by unit conversions, as we used in the previous chapters.

Important
Point

Now try the practice problems on page 177.

Taking the Test

- Wear a watch! Bring your calculator and make sure it is functioning.
- Get to class a few minutes early so you have time to get comfortable and get organized before the test is handed out.
- Put your name on the test!
- As soon as you get the test, write down on the test paper any information you are worried about forgetting. Then you won't have to worry anymore, and your mind

will be free to concentrate completely on each test question.

- Preview the entire test. It only takes a minute or two but can tell you much.

  Previewing tells you:

  the number and types of questions the test contains (multiple choice, true-false, matching, essay, problems, etc.).

  the number of points each question is worth.

  any additional information you need to write down to be sure you remember it.

  how to schedule your time.

- Previewing should also build your confidence because, if you are prepared, you will know immediately that you can answer most of the questions.

- Read any directions carefully and completely. If there is something you do not understand, raise your hand and ask about it. The instructor will probably be able to clarify it for you. In any case, the worst that can happen is the instructor will say he or she cannot answer your question.

- Begin with the questions that are easiest for you. This strategy will build your confidence and lessen your anxiety. It may also remind you of things you need to know to answer the more difficult items. In addition, it will prevent you from losing points by spending too much time on things you don't know and running out of time before completing items you do know.

- Return to questions you skipped. If they are objective questions, i.e., true-false, multiple choice, matching, etc., make an "educated guess." (See *Mastering the Multiple Choice Test*). If they are essays or problems to be solved, do as much as you can. If the problem has several dependent parts and you do not know how to do the first part but do know how to do the next part, write a note saying, "I'm not sure how to do part A, so I am going to assume the answer is (make up a reasonable value) and use this value to complete the problem." Remember, part credit is better than no credit! Answer every question as completely as possible.

- Be neat and show your work, it will be easier to check your answers.

- Check your work! If you only have a few minutes left, check the first few questions you answered. You were most nervous then, and therefore you might have made careless errors. Then check the problems you thought were the easiest. You also make careless errors when you are too relaxed. If you eliminate careless errors, you can usually increase your grade by at least 10%!

- Do not get nervous because others are leaving before you. Those who finish early often are the weakest students. Use all the time allowed. There is no prize for finishing early!

CHAPTER 10

Counting Calories
Heat and Energy

In the last chapter you learned that heat flows from something hotter to something colder in temperature. Then you learned conversions between temperature scales. But we didn't talk about what heat really is and how we measure it. That is the subject of this chapter.

What is Heat?

Heat is one form of energy. We use the word energy loosely in everyday life. For instance we say, "I'm full of energy." In science, energy is defined as the ability to do work, and work is the ability to make a change. When you carry the milk from the refrigerator to the table, you do work. When you ride your bicycle, you do work. Thinking is even work, as we all know! In a physics class you would go into the exact definitions of what work is, but in chemistry that is not necessary.

Energy can be either potential or kinetic. Potential energy is stored energy. Kinetic energy is the energy of motion. When you draw an arrow in a bow and pull back, you have potential energy stored in the tension of the bow. When you release the arrow, the potential energy changes into the kinetic energy of the moving arrow. A boulder on the edge of a cliff has potential energy that can be converted into kinetic energy if the boulder falls over the cliff.

Batteries store potential energy in the chemicals of the battery that can be converted to the kinetic energy of a moving toy. The food you eat has stored chemical potential energy that you convert to the kinetic energy of walking to class and talking about chemistry.

Energy can take many forms: the mechanical energy of motion, the electrical energy that powers your radio, the sound energy from the radio's speakers.

Although some chemical reactions produce light (we see this in the reaction of magnesium and oxygen in the old-fashioned flash bulbs) and chemical energy can be converted into electricity in batteries, in chemistry you will almost exclusively deal with heat energy.

Heat is actually a form of kinetic energy. It is the motion of molecules. The hotter something is, the faster its molecules move. On a hot summer day, the air molecules hit your skin more intensely than on a cool day. The heat you feel is all those countless, soundless collisions. In your oven the molecules are moving so fast their collisions damage the outside surface of your roast and turn it brown. That is what we call cooking.

Calories and Joules
The Units of Heat

We all experience heat. We feel the heat flow out of the oven when we open the door. We feel it flow out of our body when we walk outside on a cold day. But how do we measure heat?

The older metric unit of heat is the calorie. A calorie is defined as the amount of heat required to raise the temperature of 1 gram of water by 1 degree Celsius. For example, if we have 10 calories (cal), we can raise the temperature of 1 gram of water by $10\,°C$ or we can raise the temperature of 10 grams of water $1\,°C$. If we have 100 calories, we can raise the temperature of 1 gram of water by $100\,°C$, 10 grams of water by $10\,°C$ or 100 grams of water by $1\,°C$.

The SI unit of heat is the Joule (J). The Joule comes to us from physics and, unfortunately, cannot be as easily visualized as the calorie can. It is easy to see in your mind's eye one gram of water being heated up one degree. There is nothing really like that to visualize for the Joule.

There is, however, a conversion between calories and Joules.

$$4.184\,J = 1\,cal$$

We can add the metric prefix kilo to calories and Joules.

$$1\,kcal = 1000\,cal \qquad 1kJ = 1000\,J \qquad 4.184\,kJ = 1\,kcal$$

A kcal is the amount of heat it takes to heat 1000 grams of water $1\,°C$. Since 1000 grams is 1 kilogram, we can also say that a kcal is the heat needed to raise 1 kg of water $1\,°C$.

The kilocalorie is actually equivalent to the dietician's Calorie (written with a capital C) that you see on cereal boxes and candy bars, and in exercise books. A candy bar that has 130 Calories (130Kcal) could heat 2 kg of water (about 1/2 gal) 65°C.

We can convert between the heat units.

How many kcal is 2.42×10^4 cal?

$$2.42 \times 10^4 \text{ cal} \times \frac{1 \text{ kcal}}{1000 \text{ cal}} = 24.2 \text{ kcal}$$

How many Joules is 5219 cal? How many kJ?

We'll first convert calories to Joules to answer the first question:

$$5219 \text{ cal} \times \frac{4.184 \text{ J}}{1 \text{ cal}} = 21{,}840 \text{J or } 2.184 \times 10^4 \text{ J}$$

Now we'll convert Joules to kilojoules to answer the second question:

$$2.184 \times 10^4 \text{ J} \times \frac{1 \text{ kJ}}{1000 \text{ J}} = 21.84 \text{ kJ}$$

Aren't You BTUtiful?

You probably will never use it in your chemistry class, but you may be interested to know that we have a unit for heat in our English system. If you ever buy an air conditioner or a furnace, you will see it. It is called the British Thermal Unit or Btu.

The Btu is defined in terms of heating water just as the calorie but has English units of mass and temperature. A Btu is the heat it takes to raise the temperature of 1 pound of water 1 degree Fahrenheit. If you have 100 Btu's, you can heat 10 pounds of water 10°F. A Btu is about 1/4 of a Kilocalorie.

If an air conditioner has a sticker that rates it at 20,000 Btu's, it means it can remove that much heat from a room in 1 hour. Remember when you remove heat from something, it cools.

Specific Heat

Substances differ in how they absorb heat. If you put 1 calorie of heat into 1 gram of water, it will heat up 1°C. If you put 1 calorie of heat into 1 gram of silver, it will heat up almost 20°C. The property that relates how material absorbs heat is called specific heat. Specific heat is the amount of heat it takes to raise the temperature of 1 gram of a substance by 1 degree Celsius. If you think the definition of specific heat sounds a lot like the definition of calorie, you're right. The specific heat of water is:

$$\frac{1.0 \text{ cal}}{\text{g}\,^\circ\text{C}}$$

The specific heat of silver is:

$$\frac{0.057 \text{ cal}}{\text{g}\,^\circ\text{C}}$$

It only takes 0.057 cal (57 thousandths of a calorie) to raise the temperature of 1 gram of silver 1°C, compared to the one whole calorie to heat 1 gram of water 1°C. Table 10.1 lists the specific heat of some common substances.

Table 10.1: Specific Heats of Some Common Substances

Substance	Specific Heat (cal/g°c)
Water	1.00
Ethyl alcohol	0.58
Sand	0.19
Wood	0.42
Aluminum	0.22
Iron	0.11
Copper	0.092
Silver	0.057
Gold	0.031

The Heat Equation

We often want to know how much heat is absorbed or lost when an object is heated or cooled. The amount of heat depends on three things:

- the mass of the object (the greater the mass, the more heat an object can hold. A bathtub full of water will have more heat than a teacup of water).

- the beginning and ending temperature (it will take more heat to warm something to 80 degrees than to just 40 degrees).

- the specific heat of the material (we saw that metals absorb little heat in comparison to water).

Chapter 10

These three variables are related in a simple equation, where ΔT means "change in temperature":

$$\text{Heat (cal)} = \text{Mass(g)} \times \Delta T(°C) \times \text{Specific Heat}\left(\frac{\text{cal}}{\text{g}°\text{C}}\right)$$

Just as it is with so many equations in chemistry, if you are given any three variables, you can find the fourth. The most common problem is finding the heat if you are given mass, the temperature change and the specific heat.

Now let's try some problems.

You heat 500.0 grams (about 1 pint) of water from room temperature of 20°C to the boiling point of 100°C. How much heat does it take?

The temperature change is 80°C The specific heat of water is $\frac{1.0 \text{ cal}}{\text{g}°\text{C}}$.

$$\text{Heat} = 500.0\text{g} \times 80°\text{C} \times \frac{1.0 \text{ cal}}{\text{g}°\text{C}} = 40,000 \text{ cal} = 4.0\times10^4 \text{ cal}$$

Notice how we wrote the specific heat with a horizontal fraction bar and notice how the mass and temperature units cancel to leave us with calories.

Now suppose we have a 500 gram iron pan and we also heat it from 20°C to 100°C. How much heat does it take?
The specific heat of iron is $\frac{0.11 \text{ cal}}{\text{g}°\text{C}}$.

$$\text{Heat} = 500.0\text{g} \times 80°\text{C} \times \frac{0.11 \text{ cal}}{\text{g}°\text{C}} = 4400 \text{ cal}$$

It takes only about one-tenth the calories to heat the iron as to heat the water. Because the metal has a lower specific heat, less heat is required to increase its temperature.

If we are given heat, the temperature change, and the mass of a substance, we can calculate its specific heat.

If you divide both sides of the heat equation by mass and ΔT, we get

$$\text{Specific heat} = \frac{\text{heat}}{\text{mass} \times \Delta T}$$

This, of course, is the definition of specific heat.

What is the specific heat of magnesium if it takes 25.0 cal to heat 7.00 grams from 20°C to 27.2°C?

$$\text{Specific heat} = \frac{\text{heat}}{\text{mass} \times \Delta T} = \frac{25.0 \text{ cal}}{7.00 \text{ g} \times 7.2°C} = \frac{0.496 \text{ cal}}{\text{g}°C}$$

Looking Ahead

Later on in the course you will be given conversions between moles and heats of reaction. You will then do conversions with moles and kcal.

When methane, which is natural gas, burns, one mole gives off 213.0 kcal of heat. This is the heat you use to heat your home. If 5.25 moles of methane are burned, how many kcal are released?

This problem is just like any other unit conversion. The first sentence gives us the equality:

1 mole = 213.0 kcal

From this equality we can write two conversions:

$$\frac{213.0 \text{ kcal}}{1 \text{ mol}} \quad \text{and} \quad \frac{1 \text{ mol}}{213.0 \text{ kcal}}$$

Now we will perform the conversion:

$$5.25 \text{ mol} \times \frac{213.9 \text{ kcal}}{1 \text{ mol}} = 1120 \text{ kcal}$$

5.25 moles of methane produces 1120 kcal of heat.

Now try the practice problems on page 179.

Mastering the Multiple Choice Test

In addition to the general test-taking strategies we discussed at the end of Chapter 9, there are some specific things you can do to increase your score on multiple choice tests.

- As soon as you get the test paper, jot down any information you know you will need and are afraid you will forget.

- Scan the entire test. If any of the questions remind you of any other facts you are afraid of forgetting, write them down.

- Read the directions. Most multiple choice tests have only one correct choice, but don't assume anything.

- Answer the easiest, least time-consuming questions first. This strategy will assure that you have time to answer all the questions you know. It will also build your confidence and remind you of things that might help you answer the more difficult items.

- Before you look at the answer choices, work the problem and ask yourself "Does this answer make sense?" Remember that many of the incorrect answer choices will be derived from mistakes that students commonly make. If you see your answer as one of the choices, don't immediately mark it.

- If you don't find your answer among the choices, look again. Perhaps your answer is there but in a different form. For example 1/2 could be written as 0.5, 0.50 or 50%, etc. Maybe the answer is in scientific notation and you calculated your's in decimal form.

- After you've answered all the questions you know, carefully reread each question you skipped. You may find you now remember how to do them. Doing the questions you know warms you up and sometimes answering one question will spark your memory on how to answer another.

- If you are still unable to answer some of the questions, make an "educated" guess. Can you eliminate one or more of the choices? For example, if the question asks for grams, eliminate all the choices that don't have units of grams.

- If all else fails, simply guess. Choose a letter that is different from the question before and pick a letter that hasn't been used very often for a correct choice. Never leave a multiple choice question blank unless the directions say there is a penalty for guessing.

- Use all the time allowed. There is no prize for being the first to finish.

CHAPTER 11

Practice Tests

Chemistry tests often include multiple choice items. The Study Skills section at the end of Chapter 10 listed some important test-taking strategies for taking multiple choice tests. Now we want to give you the opportunity to review and practice these strategies.

This chapter contains two multiple choice tests. Select a quiet place such as the library and take the first ten-problem test as if you were taking it in class. Then check your work using the answer key which follows. Remember that many incorrect choices are patterned after mistakes students are known to make. So if you make an error that results in an incorrect choice, you need to determine exactly what you did wrong and to understand why your method was incorrect. Analyze any errors you made in Practice Test 1, review any concepts that gave you difficulty, and then try Practice Test 2.

Practice Test 1

_____ 1. Write 0.00102 in scientific notation.

      a. 1.0×10^{-3} b. 1.02×10^{3} c. 1.02×10^{-4} d. 1.02×10^{-3}

_____ 2. How many significant figures (sig figs) are in 4.0050?

      a. 2   b. 3   c. 4   d. 5

_____ 3. How many significant figures are in 0.002?

      a. 1   b. 2   c. 3   d. 4

_____ 4. To how many significant figures should the answer to 3.02×5.0820 be rounded?

      a. 2   b. 3   c. 5   d. 6

_____ 5. What is the correct conversion factor to change mL to L?

      a. 1000 mL/1L b. 1000L/1mL c. 1L/1000 mL d. 1L/100 mL

_____ 6. Convert 0.55 meters to centimeters.

      a. 5.5 cm   b. 55 cm   c. 550.0 cm   d. 0.0055 cm

_____ 7. Convert 27.0°C to degrees Fahrenheit.

      a. 80.6°F   b. 47.0°F   c. 106°F   d. −2.78°F

_____ 8. Mercury has a density of 13.6g/mL. How many milliliters of mercury does it take to equal 25.0 grams?

      a. 340mL   b. 0.544 mL   c. 1.84 mL   d. 2

_____ 9. Which is the smallest unit?

      a. mm   b. km   c. cm   d. m

_____ 10. If the density of gold is 19.32 g/cm^3, what is its specific gravity?

      a. 1.932×10^{1} g/cm^3 b. 19   c. 19.32g/mL d. 19.32

Practice Test 2

_____1. Write 108,000 in scientific notation.

   a. 108×10^3 b. 1.08×10^5 c. 10.8×10^{-4} d. 108×10^{-3}

_____2. How many significant figures are in 0.0105?

   a. 2   b. 3   c. 4   d. 5

_____3. To how many significant figures should the answer to 1.04×0.02 be rounded?

   a. 1   b. 2   c. 3   d. 4

_____4. What is the correct conversion factor to change g to mg?

   a. 100 mg/1 g b. 1 g/1000 mg c. 1 mg/1000 g d. 1000 mg/1 g

_____5. Convert 5.5 mL to L.

   a. 5500 L   b. 0.055 L   c. 0.0055 L   d. 0.55 L

_____6. Convert 2.1 km to cm.

   a. 2.1×10^{-5} cm b. 21×10^4 cm c. 2.1×10^5 cm d. 2.10×10^4 cm

_____7. Convert 60°F to degrees Celsius.

   a. 15.6°C   b. 1.33°C   c. 140°C   d. 42.2°C

_____8. The density of mercury is 13.6 g/mL. What is its specific gravity?

   a. 14   b. 13.6   c. 13.6 g/cm^3 d. 1.36×10^1 g/mL

_____9. Gold has a density of 19.32 g/cm^3. How many grams of gold does it take to equal 16.4 cm^3 (about 1 cubic inch) of gold?

   a. 1.18 g   b. 317 g   c. 0.849 g   d. 22.9 g

_____10. How many calories does it take to raise the temperature of 44.9 g of gold from 22.0°C to 57.0°C? The specific heat of gold is 0.0308 cal/g °C.

   a. 34.6 cal   b. 48.4 cal   c. 30.4 cal   d. 78.8 cal

Chapter 11

Practice Test 1 Answer Key

1. **d** Move the decimal point until you get a coefficient between 1 and 10 then attach the appropriate power of 10. (Or use the appropriate keys on the calculator.)
2. **d** The 4, 5 and all the zeros are significant.
3. **a** All the zeros are place holders and not significant.
4. **b** The product can have no more sig figs than the smallest number of sig figs in any of the factors. Note the problem doesn't ask you to calculate the answer.
5. **c** $mL \times \dfrac{L}{mL}$ = L Set up the conversion so the units cancel correctly then insert the correct numbers. Notice it is easier to see the relationship if we write the conversion with a horizontal fraction line.
6. **b** $0.55\,m \times \dfrac{100\,cm}{1\,m}$ = 55 cm (2 sig figs)
7. **a** °F = 1.8(27.0) + 32 = 80.6°F (3 sig figs)
8. **c** $25.0g \times \dfrac{1\,mL}{13.6\,g}$ = 1.84mL (3 sig figs)
9. **a** Remember to relate the metric units to sizes with which you are familiar.
10. **d** Specific gravity is the density without the units.

Practice Test 2 Answer Key

1. **b** Move the decimal point until you get a coefficient between 1 and 10 then attach the appropriate power of 10. (Or use the appropriate keys on the calculator.)
2. **b** The 1, 5 and the 0 between them are significant. The two leading zeros are place holders.
3. **a** The product can have no more sig figs than the smallest number of sig figs in any of the factors. Note the problem doesn't ask you to calculate the answer.
4. **d** $g \times \dfrac{mg}{g}$ Set up the conversion so the units cancel correctly then insert the correct numbers.
5. **c** $5.5\,mL \times \dfrac{1\,L}{1000\,mL}$ = 0.0055L (2 sig figs)
6. **c** $2.1\,km \times \dfrac{1000\,m}{1\,km} \times \dfrac{100\,cm}{1\,m}$ = 210,000cm = 2.1×10^5 cm (2 sig figs)
7. **a** °C = (60.0 − 32)÷1.8 = 15.6°C (3 sig figs)
8. **b** Specific gravity is the density without the units.
9. **b** $16.4\,cm^3 \times \dfrac{19.32\,g}{1\,cm^3}$ = 317 g (3 sig figs)
10. **b** 44.9g × 35°C × 0.0308cal/g°C = 48.4 cal (3 sig figs)

Furry Little Unit Conversions
Moles

By this time in the term you've studied several chapters in your chemistry text that did not include conversions. In those chapters you looked at the structure of atoms. You studied how atoms are composed of protons and neutrons in the nucleus with electrons in orbitals surrounding the nucleus. You studied how atoms can gain and lose electrons from their outermost levels to form ions and how atoms can share electrons to form covalent bonds. You also learned to name simple ionic and molecular compounds.

However, individual atoms and molecules are much too small to work with routinely in a laboratory or in everyday life. That is why we need the unit that is the subject of this chapter, the mole.

Mountains and Mole Hills
What is a Mole?

A mole (abbreviated mol) is a defined number of things, just as a pair is two, a dozen is twelve or a gross is 144. The only difference is that a mole is a very large number of things: 6.02×10^{23} items to be exact. This number, which will soon be burned into your memory, is a billion billions and then 602 thousands of those billion billions. We call it Avogadro's number. This number is unimaginably big, yet you can swallow Avogadro's number of water molecules in one small gulp.

The mole is important in chemistry because, although we cannot easily keep track of a single carbon atom, we can hold a mole of carbon atoms in the palm of our hand. While we cannot see one atom of iron react with one atom of sulfur, we can watch 1 mole of iron (6.02×10^{23} iron atoms) react with 1 mole of sulfur (6.02×10^{23} sulfur atoms).

If a choreographer had a dozen female dancers and wanted to have one male dancer for every female, he would need to hire a dozen male dancers. If the choreographer wanted two male dancers for every female, he would need two dozen male dancers. Likewise, if we want to react one molecule of hydrogen with every one molecule of oxygen, then we mix one mole of hydrogen with one mole of oxygen. If we want to react two molecules of hydrogen to every one molecule of oxygen, then we mix two moles of hydrogen with one mole of oxygen.

Since a mole is simply a number, that is 6.02×10^{23} items, we can have a mole of anything. We can have a mole of marbles, a mole of baseballs or a mole of elephants. In chemistry we deal with moles of microscopic particles. A mole of carbon, a mole of iron, and a mole of silicon each contains 6.02×10^{23} atoms. A mole of the diatomic elements nitrogen (N_2) and oxygen (O_2) each contains 6.02×10^{23} molecules. A mole of the covalent compounds glucose and methane each contains 6.02×10^{23} molecules. A mole of the ionic compounds KCl and NaF each contains 6.02×10^{23} formula units. It doesn't matter whether we are talking about atoms, molecules or formula units, Avogadro's number is always the same.

By the way, you may have wondered where this funny name "mole" came from. It comes from the Latin word which means heap or pile. A mole is literally a heap, just like the furry little animal, the mole, makes heaps we call mole hills in your lawn.

Finding the Molar Mass

We can not count out a mole of atoms or molecules, but it is very easy to weigh a mole. The mass of a mole of a substance is its formula weight expressed in grams and is called the molar mass. Let's look at some examples. The formula weight of water (H_2O) is 18 amu. Therefore a mole of water has a mass of 18 grams. Glucose ($C_6H_{12}O_6$) has a formula weight of 180 amu. Therefore a mole of glucose has a mass of 180 grams. A glucose molecule is ten times heavier than a water molecule. So Avogadro's number of glucose molecules will weigh ten times more than Avogadro's number of water molecules (180g/18g = 10).

Just like a dozen eggs has a dozen yolks and a dozen whites, we can have moles within moles. One mole of carbon monoxide (CO) has 1 mole of carbon atoms and 1 mole of oxygen atoms. One mole of glucose ($C_6H_{12}O_6$) has 6 moles of carbon atoms, 12 moles of hydrogen atoms and 6 moles of oxygen atoms.

To calculate the molar mass, we add up the masses of all the atoms in the formula.

Let's calculate the molar mass of aluminum sulfate, $Al_2(SO_4)_3$:

2 moles of Al atoms	$2 \times 27.0 =$	54.0 g
3 moles of S atoms	$3 \times 32.1 =$	96.3 g
12 moles of O atoms	$12 \times 16.0 =$	192.0 g
		342.3 g

So we know that a mole of aluminum sulfate has a mass of 342.3 grams. Or we can say that 6.02×10^{23} formula units of aluminum sulfate weigh 342.3 grams.

What is the molar mass of sucrose? The molecular formula of sucrose is $C_{12}H_{22}O_{11}$.

12 moles of C atoms	12×12.0 gram $=$	144 g
22 moles of H atoms	22×1.0 gram $=$	22 g
11 moles of O atoms	11×16.0 gram $=$	176 g
		342 g

So we can write the relationship

$$1 \text{ mole } C_{12}H_{22}O_{11} = 342 \text{ grams } C_{12}H_{22}O_{11}.$$

Between Moles and Grams
The Most Important Mole Coversions

We can derive unit conversions for moles just like we did before for other units. We just saw that for sucrose, 1 mole $C_{12}H_{22}O_{11} = 342$ grams. From this relationship we can derive two conversions.

$$\frac{342 \text{ gram sucrose}}{1 \text{ mole sucrose}} \quad \text{or} \quad \frac{1 \text{ mole sucrose}}{342 \text{ gram sucrose}}$$

If we want to convert from moles to grams, we use the first conversion. If we want to convert from grams to moles, we use the second.

This type of conversion between moles and grams is one of the most routine that people who work in a chemistry laboratory do. Now let's do some conversions between moles and grams.

What is the weight in grams of 0.25 mole of sucrose?

We use the conversion $\dfrac{342 \text{ gram sucrose}}{1 \text{ mole sucrose}}$ so that mole units cancel:

$$0.25 \text{ mol sucrose} \times \frac{342 \text{ g sucrose}}{1 \text{ mol sucrose}} = 86 \text{ g sucrose}$$

How many moles are 125 g sucrose?

Now we use the conversion $\dfrac{1 \text{ mole sucrose}}{342 \text{ grams sucrose}}$ so that the gram units cancel:

$$125 \text{ g sucrose} \times \frac{1 \text{ mol sucrose}}{342 \text{ g sucrose}} = 0.365 \text{ mol sucrose}$$

These two conversions involving sucrose demonstrate the two types of conversions you can do using molar mass. Now let's do a few more.

Geraniol, partially responsible for the scent of roses, has a formula of $C_{10}H_{18}O$. What is the mass of 2.50 moles of geraniol?

We first calculate the molar mass.

10 moles of C atoms	$10 \times 12.0 \text{ g} = 120 \text{ g}$
18 moles of H atoms	$18 \times 1.0 \text{ g} = 18 \text{ g}$
1 mole of O atoms	$1 \times 16.0 \text{ g} = \underline{16.0 \text{ g}}$
	154 g

So we can write two conversion factors:

$$\frac{154 \text{ g geraniol}}{1 \text{ mole geraniol}} \quad \text{or} \quad \frac{1 \text{ mole geraniol}}{154 \text{ g geraniol}}$$

We choose the first so that the units cancel.

$$2.50 \text{ mol geraniol} \times \frac{154 \text{ g geraniol}}{1 \text{ mol geraniol}} = 385 \text{ g geraniol}$$

How many moles are there in 58.0 grams of toluene, C_7H_8?

First you calculate the molar mass of toluene from the molecular formula.

7 moles of C atoms	$7 \times 12.0 \text{ g} = 84.0 \text{ g}$
8 moles of H atoms	$8 \times 1.0 \text{ g} = \underline{8.0 \text{ g}}$
	92.0 g

Then you do the conversion. First write down the given quantity and then multiply by the conversion that cancels the gram units.

$$57.0 \cancel{\text{ g toluene}} \times \frac{1 \text{ mol toluene}}{92.0 \cancel{\text{ g toluene}}} = 0.62 \text{ mol toluene}$$

How much does 7.2 moles of methane CH_4 weigh?

The molar mass of methane is 16 grams.

$$7.2 \cancel{\text{ mol } CH_4} \times \frac{16.0 \text{ g } CH_4}{1 \cancel{\text{ mol } CH_4}} = 115 \text{ g } CH_4$$

In all these conversions we used the molar mass that we calculated from the molecular formula to convert back and forth between moles and grams.

A tablet of an antacid contains 500.0 milligrams of aluminum hydroxide which has a formula of $Al(OH)_3$. How many moles of $Al(OH)_3$ are in one tablet?

Since we have not been given the molar mass, we need to calculate it:

1 mole of Al atoms	$1 \times 27.0 \text{ g} = 27.0 \text{ g}$
3 moles of O atoms	$3 \times 16.0 \text{ g} = 48.0 \text{ g}$
3 moles of H atoms	$3 \times 1.0 \text{ g} = \underline{3.0 \text{ g}}$
	78.0 g

Next we write down our given quantity, 500.0 mg $Al(OH)_3$. First we convert mg to grams and then grams to moles.

$$500.0 \cancel{\text{ mg } Al(OH)_3} \times \frac{1 \text{ g}}{1000 \text{ mg}} \times \frac{1 \text{ mol } Al(OH)_3}{78.0 \cancel{\text{ g } Al(OH)_3}} = 0.00641 \text{ mol } Al(OH)_3$$

In scientific notation the answer is 6.41×10^{-3} mol $Al(OH)_3$.

Counting Atoms
Conversions With Avogadro's Number

Conversions with Avogadro's number are easy. You only need to know that one special number, 6.02×10^{23}. The only difference is the number of "things" you are talking about.

For the element iron

$$1 \text{ mole Fe} = 6.02 \times 10^{23} \text{ Fe atoms}$$

For the diatomic element nitrogen

$$1 \text{ mole } N_2 = 6.02 \times 10^{23} \text{ } N_2 \text{ molecules}$$

For the molecular compound formaldehyde, CH_2O,

$$1 \text{ mole } CH_2O = 6.02 \times 10^{23} \text{ } CH_2O \text{ molecules}$$

For the ionic compound potassium chloride, KCl

$$1 \text{ mole KCl} = 6.02 \times 10^{23} \text{ KCl formula units}$$

It does not matter whether you are dealing with atoms, molecules, or formula units, the calculation is the same.

Now let's use Avogadro's number to do some conversions between moles and numbers of particles.

How many molecules are there in 0.75 moles of ethyl alcohol, C_2H_6O?

$$0.75 \text{ mol } C_2H_6O \times \frac{6.02 \times 10^{23} \text{ molecules}}{1 \text{ mol } C_2H_6O} = 4.5 \times 10^{23} \text{ molecules}$$

How many moles is one billion (1.0×10^9) molecules of water?

$$1.0 \times 10^9 \text{ molecules } H_2O \times \frac{1 \text{ mole } H_2O}{6.02 \times 10^{23} \text{ molecules } H_2O} = 1.66 \times 10^{-15} \text{ mole } H_2O$$

A billion water molecules is far less than a trillionth (10^{-12}) of a mole.

Now try the practice problems on page 181.

CHAPTER 13

Know Any Good Recipes? Stoichiometry

When you make pancakes or bake a loaf of bread, you probably use a recipe. A recipe is a set of instructions that tells you how much of each ingredient to mix together to obtain the final product—the pancakes or the bread.

Here is a very simple recipe for pancakes.

$$\text{1 cup flour} + \text{3 eggs} + \text{2 cups milk} = \text{9 pancakes}$$

This "equation" tells us that if we want to make 9 pancakes, we mix 1 cup of flour with three eggs and 2 cups of milk.

We can double the recipe if we want. Then we would mix 2 cups of flour with 6 eggs and 4 cups of milk, and make 18 pancakes.

A chemical equation is also a recipe. Using a chemical equation as a recipe is called Sto-ich-io-metry. It tells the chemist how much of each reactant to mix together and how much of the product she will produce. Let's look at the equation a chemist will use if she wants to make ammonia (NH_3) from the elements hydrogen (H_2) and nitrogen (N_2).

$$3H_2 + N_2 \rightarrow 2NH_3$$

Just like the pancake recipe, this equation tells the chemist that if she reacts 3 moles of H_2 with 1 mole of N_2, she will form 2 moles of NH_3. And just as with the pancake recipe, she can double the recipe if she chooses.

From Molecules to Moles

Now let's look at the equation for making ammonia in a little more detail. You learned that a chemical equation describes a chemical reaction and a chemical reaction is simply a rearrangement of atoms. The equation

$$3\,H_2 \;+\; N_2 \;\rightarrow\; 2\,NH_3$$

at the molecule level means

3 molecules H_2 + 1 molecule N_2 → 2 molecules NH_3

If you drew pictures of the molecules in the equation, they would look like this.

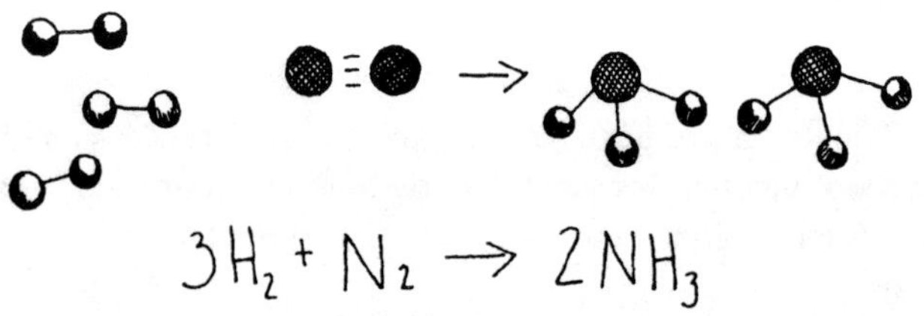

$$3H_2 + N_2 \rightarrow 2NH_3$$

Figure 13.1

The three hydrogen molecules rearrange with the nitrogen molecule to form two ammonia molecules. The equation is balanced because we have six hydrogen atoms on each side and two nitrogen atoms on each side.

If we double the coefficients on each side, we get

$$6\,H_2 \;+\; 2\,N_2 \;\rightarrow\; 4\,NH_3$$

We can multiply the coefficients on each side by 10 and get:

$$30H_2 + 10N_2 \;\rightarrow\; 20NH_3$$

or by a billion (1×10^9) and get:

$$3\times10^9\,H_2 \;+\; 1\times10^9\,N_2 \;\rightarrow\; 2\times10^9\,NH_3$$

We can even multiply by Avogadro's number (6.02×10^{23})

$$3(6.02 \times 10^{23})H_2 + 1(6.02 \times 10^{23})N_2 \rightarrow 2(6.02 \times 10^{23})NH_3$$

Remember 6.02×10^{23} molecules is one mole, so we can write

$$3 \text{ moles } H_2 + 1 \text{ mole } N_2 \rightarrow 2 \text{ moles } NH_3$$

So we can look at a chemical equation two ways: on the small scale in terms of molecules

$$3 \text{ molecules } H_2 + 1 \text{ molecule } N_2 \rightarrow 2 \text{ molecules } NH_3$$

or on the large scale in terms of moles.

$$3 \text{ moles } H_2 + 1 \text{ mole } N_2 \rightarrow 2 \text{ moles } NH_3$$

The mole interpretation is the most useful if you want to use the equation as a recipe to make chemical compounds such as ammonia.

We can derive unit conversions from a balanced chemical equation. We can see in the ammonia equation

$$3H_2 + N_2 \rightarrow 2NH_3$$

that for every 1 mole of N_2 that reacts, 3 moles of H_2 also react. So we can write two conversions

$$\frac{3 \text{ mole } H_2}{1 \text{ mole } N_2} \quad \text{and} \quad \frac{1 \text{ mole } N_2}{3 \text{ mole } H_2}$$

The second conversion is the reciprocal of the first. These conversions are called mole ratios.

We can make the same ratios for our pancake recipe:

$$1 \text{ cup flour} + 3 \text{ eggs} + 2 \text{ cups milk} = 9 \text{ pancakes}$$

For instance, for the relation between flour and eggs we can write:

$$\frac{1 \text{ cup flour}}{3 \text{ eggs}} \quad \text{and} \quad \frac{3 \text{ eggs}}{1 \text{ cup flour}}$$

We can write conversions for the relationship between N_2 and NH_3:

$$\frac{2 \text{ mole } NH_3}{1 \text{ mole } N_2} \quad \text{and} \quad \frac{1 \text{ mole } N_2}{2 \text{ mole } NH_3}$$

And for the relationship between H_2 and NH_3:

$$\frac{2 \text{ mole } NH_3}{3 \text{ mole } H_2} \quad \text{and} \quad \frac{3 \text{ mole } H_2}{2 \text{ mole } NH_3}$$

Now let's try some problems

From Moles to Moles

If 6 moles of N_2 are reacted, how many moles of NH_3 are produced?

$$3\,H_2 \; + \; N_2 \; \rightarrow \; 2\,NH_3$$

Use the ratio of the coefficients of N_2 to NH_3 to set up the conversion.

$$6\,\text{mol } N_2 \times \frac{2\,\text{mol } NH_3}{1\,\text{mol } N_2} = 12\,\text{mol } NH_3$$

If 30 moles of H_2 are reacted, how many moles of N_2 also react?

$$3\,H_2 \; + \; N_2 \; \rightarrow \; 2\,NH_3$$

Use the ratio of the coefficients of H_2 to N_2 and set up the conversion:

$$30\,\text{mol } H_2 \times \frac{1\,\text{mol } N_2}{3\,\text{mol } H_2} = 10\,\text{mol } N_2$$

If 6 moles of butane, C_4H_{10}, are burned according to the following balanced equation, how many moles of oxygen, O_2, also react?

$$2\,C_4H_{10} \; + \; 13\,O_2 \; \rightarrow \; 8\,CO_2 \; + \; 10\,H_2O$$

Note that only the ratio of C_4H_{10} to O_2 concerns us in this problem. We can put marks over these compounds in the equation to remind us that, in this problem, it is only C_4H_{10} and O_2 that concern us. The equation tells us that for every 2 moles of C_4H_{10} that react, 13 moles of O_2 also react.

$$6\,\text{mol } C_4H_{10} \times \frac{13\,\text{mol } O_2}{2\,\text{mol } C_4H_{10}} = 39\,\text{mol } O_2$$

From Grams to Grams

Remember that we cannot measure moles directly. If we want to measure out a certain number of moles, we have to convert to grams. (You may want to review mole conversions in the last chapter.)

This takes us to the second type of stoichiometry problem in which we are given grams of one reactant or product and want to find the grams of another reactant or product.

The road map we will use looks like this:

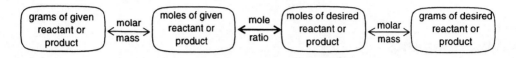

The only difference between these problems and the mole-to-mole problems we just did in the previous section is the addition of two gram-to-mole conversions.

Let's try some problems. How many grams of Al are produced if 50.0 grams of Al_2O_3 are reacted according to the following balanced equation?

$$2\,Al_2O_3 \rightarrow 4\,Al + 3\,O_2$$

Remember it's always moles that react with moles. If you are given grams in a stoichiomtry problem, you always have to convert to moles. We first need to calculate the molar mass of Al_2O_3:

$$
\begin{array}{ll}
\text{2 mole Al} & \text{2} \times \text{27.0g} = \text{54.0g} \\
\text{3 mole O} & \text{3} \times \text{16.0g} = \underline{\text{48.0g}} \\
& \hspace{2.5em} \text{102.0g}
\end{array}
$$

The chemical equation tells us that for every 2 moles of Al_2O_3 that react, 4 moles of Al are produced. Now let's set up the conversion.

$$\overset{\checkmark}{2\,Al_2O_3} \rightarrow \overset{\checkmark}{4\,Al} + 3\,O_2$$

$$50.0\,\text{g}\,Al_2O_3 \times \frac{1\,\text{mol}\,Al_2O_3}{102.0\,\text{g}\,Al_2O_3} \times \frac{4\,\text{mol}\,Al}{2\,\text{mol}\,Al_2O_3} \times \frac{27.0\,\text{g}\,Al}{1\,\text{mol}\,Al} = 26.5\,\text{g}\,Al$$

In stoiciometry problems always be sure to carefully label your units!

If 125 grams of Al_2O_3 react, how many grams of O_2 are produced?

$$\overset{\checkmark}{2\,Al_2O_3} \rightarrow 4\,Al + \overset{\checkmark}{3\,O_2}$$

The equation tells us that for every 2 moles of Al_2O_3 that react, 3 moles of O_2 are produced. The molar mass of O_2 is 32.0 g. Now set up the conversion.

$$125\,g\,Al_2O_3 \times \frac{1\,mol\,Al_2O_3}{102.0\,g\,Al_2O_3} \times \frac{3\,mol\,O_2}{2\,mol\,Al_2O_3} \times \frac{32.0\,g\,O_2}{1\,mol\,O_2} = 58.8\,g\,O_2$$

Limiting Reagents

Let's look back at our pancake recipe.

Suppose you want to prepare a large batch of pancakes and when you go to your pantry, you find a full bag of flour (10 cups) and a gallon (8 cups) of milk, but only 6 eggs. How many pancakes can you prepare from these ingredients? Let's see how many pancakes you could make using all of each ingredient.

$$10\,cups\,flour \times \frac{9\,pancakes}{1\,cup\,flour} = 90\,pancakes$$

$$6\,eggs \times \frac{9\,pancakes}{3\,eggs} = 18\,pancakes$$

$$8\,cups\,milk \times \frac{9\,pancakes}{2\,cups\,milk} = 36\,pancakes$$

We will clearly run out of eggs before we run out of milk or flour so we speak of eggs as the "limiting reagent"—the substance that will be used up first in the reaction.

You can also think of limiting reagents as being like a dance. Imagine that there are 13 men and 20 women at a dance. There can only be 13 male/female couples dancing at any one time. The men in this case are the "limiting reagents."

Have you ever seen what happens when you put a glass over a burning candle? It soon sputters out. The candle burns by reacting with oxygen in the air. Oxygen is not normally a limiting reagent because it is so plentiful in the air. But as soon as you put a glass over the candle, the flame is limited to the oxygen trapped inside. As soon as that's used up, the flame sputters out.

Now try the practice problems on page 183.

CHAPTER 14

A Lot of Hot Air
Gas Laws

One of the topics in chemistry involving the most use of algebra is the properties of gases. You might want to review the material on proportions in Chapter 7 before continuing with this chapter.

The Properties of Gases

Gases have four important properties: pressure (P), volume (V), temperature (T), and the amount of moles of gas (n). Let's look at each of these properties in detail.

Pressure

A gas exerts a pressure on all the surfaces of its container. The pressure of the air in your car tires keeps the tires firm and inflated. If one of your tires gets punctured and the gas escapes, you lose pressure and your tire goes flat. Pressure is defined as force per area:

$$\text{Pressure} = \frac{\text{force}}{\text{area}}$$

In the English system we use the units of pounds per square inch, which you may have seen abbreviated as "psi." If you look at the small print on your car tires, you will find written something such as "inflate to 35 psi."

The atmospheric air exerts pressure on every surface on earth. This pressure is literally the weight of the air. The average air pressure at sea level is $\dfrac{14.7 \text{ pounds}}{1 \text{ inch}^2}$ = 14.7 psi. The air pressure around us constantly changes as high pressure and low pressure fronts move into an area. Pressure also decreases as you go up in altitude. At 18,000 feet above sea level the atmospheric pressure decreases to 7.3 psi.

Instead of using pressure units of force per area, in chemistry we work with the effect the pressure has on a mercury barometer. To make a mercury barometer, you prepare a tube at least 800 mm long (about 32 inches) and with one end closed and fill it with mercury, being careful to prevent any air bubbles from being trapped. Then you carefully turn it upside down into a dish of mercury. You will find that not all the mercury runs out, but a column about 760 mm long is left with an empty space at the closed end. Standard atmospheric pressure pushes the mercury 760 mm up the tube. From this measurement we can define a standard atmosphere, (abbreviated atm).

$$1 \text{ atm} = 760 \text{ mm Hg}$$

A millimeter of mercury is also called a torr after Evangelista Torricelli, who invented the barometer. So we can also say

$$1 \text{ atm} = 760 \text{ mm Hg} = 760 \text{ torr}$$

In the United States we measure pressure in inches of mercury instead of millimeters.

$$760 \text{ mm} = 29.9 \text{ inches}$$

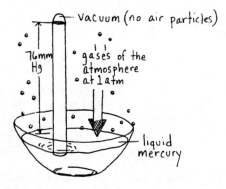

Figure 14.1

If you look at the fine print of the weather report in your newspaper, you will find the barometric pressure reported in inches of mercury. If you follow the pressure where you live, you will find that the height, measured in inches of mercury, is constantly changing; it falls before storms come and rises before fair weather comes.

The SI unit of pressure is the pascal (Pa). 1 atm = 101,325 Pa. The pascal is so small that it is often converted to kilopascals. 1 atm = 101,325 Pa = 101.325 kPa. Table 14.1 lists all the common pressure units. The units in bold are the ones you will use most often in chemistry problems.

Volume

There is nothing new to say about the volume units used with gases. You will use both liters (L) and milliliters (mL).

Temperature

When you do calculations with gas laws, you will always use the Kelvin scale. In Chapter 9, we looked at the relationship between Celsius degrees and Kelvins.

$$K = {}^\circ C + 273$$

In fact the main reason for learning and using the Kelvin scale is so that you can work with it in gas law problems. When you work with gas volumes, you need an absolute scale that starts at zero with no negative temperatures.

Table 14.1: Units for Measuring Pressure

Unit	Abbreviation	Unit Equivalent to 1 atm
Atmosphere	**atm**	**1 atm**
Millimeters of Hg	**mm Hg**	**760 mm Hg**
Torr	**torr**	**760 torr**
Inches of Hg	in. Hg	29.9 in. Hg
Pounds per sq inch	psi	14.7lb/in^2
Pascal	Pa	101,325Pa

Amount of Gas

The amount of gas in a container affects its properties. An aerosol can, for instance, has a high pressure when it is full of gas. Its pressure decreases as you let gas out of it. That is, when there are more particles in a container, the pressure is higher than when there are fewer particles. The amount of gas is usually measured in moles and is symbolized as n.

The Combined Gas Law

You are going to study three laws that relate the pressure, volume and temperature of a gas. The relationships can be summarized in one equation called the combined gas law:

$$\frac{P_1 V_1}{T_1} = \frac{P_2 V_2}{T_2}$$

P_1 is the initial pressure P_2 is the final pressure

V_1 is the initial volume V_2 is the final volume

T_1 is the initial temperature T_2 is the final temperature

We will look at each of these relationships in detail in the next sections.

Pistons and Plungers
Boyle's Law

If you decrease the volume of a fixed amount of gas at a constant temperature, the pressure of the gas increases. You can experience this effect if you plug the nozzle of a bicycle pump with your finger. As you push down on the plunger, you feel more resistance because the pressure increases as you squeeze the air in the pump into a smaller volume. Figure 14.2 illustrates this. This effect is called Boyle's law.

Boyle's law is an example of an inverse relationship. Pressure varies inversely with volume at a constant temperature. As the volume of a gas increases, the pressure has to decrease. As the volume of a gas decreases, the pressure has to increase. We can state Boyle's law mathematically:

$$P = \frac{\text{constant}}{V} \text{ or } PV = \text{constant}$$

or as

$$P_1 V_1 = P_2 V_2$$

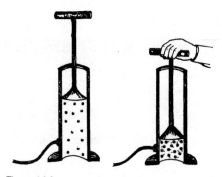

Figure 14.2

We can derive Boyle's law from the combined gas law by crossing out the constant temperature.

$$\frac{P_1 V_1}{\cancel{T_1}} = \frac{P_2 V_2}{\cancel{T_2}} \quad \text{at constant temperature}$$

Let's try a few problems involving Boyle's law.

If 8.0 liters of methane gas at normal atmospheric pressure (1.0 atm) is compressed into a volume of 2.0 liters at constant temperature, what is the new pressure?

List the variables:

$P_1 = 1.0$ atm $\qquad\qquad P_2 = ?$

$V_1 = 8.0$ L $\qquad\qquad V_2 = 2.0$ L

Before we solve, look at the numbers and see how the pressure must change. Since the volume is decreasing, the pressure must increase. We solve for P_2:

$$P_2 = P_1 \times \frac{V_1}{V_2}$$

$$P_2 = 1.0 \text{ atm} \times \frac{8.0 \cancel{L}}{2.0 \cancel{L}} = 4.0 \text{ atm}$$

If the pressure gauge on a 40.0 L tank of helium reads a pressure of 136 atm, what is the total volume of gas the tank can deliver at 1.0 atm.

List the variables:

$P_1 = 136$ atm $\qquad\qquad P_2 = 1.0$ atm

$V_1 = 40.0$ L $\qquad\qquad V_2 = ?$

Since the pressure is decreasing, the volume must increase.

Now solve for V_2:

$$V_2 = V_1 \times \frac{P_1}{P_2}$$

$$V_2 = 40.0 \text{ L} \times \frac{136 \text{ atm}}{1.0 \text{ atm}} = 5440 \text{ L}$$

You can see that one small tank of compressed helium can fill a lot of balloons.

Hot Air Balloons
Charles' Law

If you heat a fixed amount of gas at a constant pressure, the volume of the gas increases. This direct relationship is called Charles' law. If you have seen a hot air balloon being filled, you have seen an example of Charles' law. As the air inside the balloon is heated, it expands to fill the volume of the balloon. For the same reason, air that is heated by warm ground rises as it expands in volume, becomes less dense, and floats on the colder air. You can see hawks riding on the rising, expanded hot air masses called thermals.

We state Charles' law mathematically as

$$V = \text{constant} \times T \qquad \text{or} \qquad \frac{V}{T} = \text{a constant}$$

from which we can derive:

$$\frac{V_1}{T_1} = \frac{V_2}{T_2}$$

We can derive Charles' law from the combined gas law by crossing out the constant pressure.

$$\frac{\cancel{P_1} V_1}{T_1} = \frac{\cancel{P_2} V_2}{T_2}$$

In Charles' law problems, the constant pressure is usually the atmospheric pressure where the experiment is being performed. In problems, pressure will either not be mentioned or it will be stated that pressure is constant.

A 250 mL sample of air is heated from 20°C to 400.0°C at a constant atmospheric pressure. What is the new volume of the gas?

First we have to convert the Celsius temperatures to Kelvins.

$$T_1 = 20 + 273 = 293\,K \qquad T_2 = 400 + 273 = 673\,K$$

Now we can summarize our information:

$V_1 = 250$ mL $\qquad\qquad\qquad V_2 = ?$

$T_1 = 293$ K $\qquad\qquad\qquad T_2 = 673$ K

Since the temperature is increasing, the volume must also increase. We solve for V_2:

$$V_2 = V_1 \times \frac{T_2}{T_1}$$

Then plug in the numbers.

$$V_2 = 250\text{ mL} \times \frac{673\text{ K}}{293\text{ K}} = 574\text{ mL}$$

Gay-Lussac's Law

If we keep volume constant in the combined gas law we get a new law.

$$\frac{P_1 V_1}{T_1} = \frac{P_2 V_2}{T_2}$$

$$\frac{P_1}{T_1} = \frac{P_2}{T_2}$$

This law says that if we heat a fixed quantity of gas at a constant volume, the pressure must increase. Pressure is directly proportional to temperature under these conditions. This is also the reason that aerosol cans carry the warning "Do not store near heat." If the can is heated, the pressure inside can increase so much that it would explode. The pressure in the relatively constant volume of your car tires is higher when you drive long hours on a hot summer day than it is in the morning when the tires are cool.

If you heat 250 mL of gas at 1.0 atm from 0°C to 200°C, keeping the volume constant, what is the pressure of the gas?

First we convert the Celsius temperatures to Kelvin.

$$T_1 = 0 + 273 = 273 \text{ K} \qquad T_2 = 200 + 273 = 473 \text{ K}$$

We summarize the values:

$P_1 = 1.0 \text{ atm}$ $\qquad\qquad P_2 = ?$

$T_1 = 273 \text{ K}$ $\qquad\qquad T_2 = 473 \text{ K}$

Since temperature is increasing, the pressure has to increase. We solve for P_2:

$$P_2 = P_1 \times \frac{T_2}{T_1}$$

Now we plug in the numbers from the list:

$$P_2 = 1 \text{ atm} \times \frac{473 \text{ K}}{273 \text{ K}} = 1.7 \text{ atm}$$

Notice that as the temperature almost doubled, the pressure of the gas also almost doubled.

The Combined Gas Law Revisited

It is possible that pressure, volume, and temperature can all change. If this happens we use the whole combined gas law. Let's try a problem that does this.

A helium balloon on the ground where the pressure is 765 mm Hg and the temperature is 25 °C has a volume of 2.0 liters. The balloon is released and floats up to an altitude where the pressure is 660 mm Hg and the temperature is 18 °C. What is the volume of the balloon at this altitude?

First we convert the Celsius temperatures to Kelvins.

$$T_1 = 25 + 273 = 298 \text{ K} \qquad T_2 = 18 + 273 = 291 \text{ K}$$

We summarize the data

$P_1 = 765 \text{ mm Hg}$ $\qquad\qquad P_2 = 660 \text{ mm Hg}$

$V_1 = 2.0 \text{ L}$ $\qquad\qquad V_2 = ?$

$T_1 = 298 \text{ K}$ $\qquad\qquad T_2 = 291\text{K}$

Before we plug the numbers in the equation, let's look at the effects of the pressure and temperature changes individually.

The decreased pressure, by itself, will cause the volume of the balloon to increase (Boyle's law). The decreased temperature, by itself, will cause the volume to decrease (Charles' law). Now let's solve for V_2.

$$\frac{P_1 V_1}{T_1} = \frac{P_2 V_2}{T_2}$$

$$V_2 = \frac{V_1 P_1 T_2}{P_2 T_1} = V_1 \times \frac{P_1}{P_2} \times \frac{T_2}{T_1}$$

Now we plug in the numbers

$$V_2 = 2.0 \text{ L} \times \frac{765 \text{ mm Hg}}{660 \text{ mm Hg}} \times \frac{291 \text{ K}}{298 \text{ K}} = 2.26 \text{ L}$$

The decrease in pressure was larger than the decrease is temperature, so the balloon increased in volume.

Blowing Up Balloons
Avogadro's Law

In all the examples we have looked at so far, we assumed that the amount of the gas remained constant. We will now consider what happens when we add or take away gas. When you blow up a balloon, you increase its volume by blowing more air molecules into it. If your tire gets a leak, its volume decreases as air leaks out. These are examples of Avogadro's law which says the volume of a gas is directly proportional to the number of moles. If we double the number of moles of gas, at a constant temperature and pressure, we will double the volume.

We can state this mathematically, just like the other direct proportion gas laws, as

$$\frac{V_1}{n_1} = \frac{V_2}{n_2}$$

A friend inflates a balloon to 2 liters by blowing 0.2 moles of air into it. You volunteer to blow it up some more and blow in another 0.3 moles of air to make a total of 0.5 moles. What is the new volume of the balloon? Assume the atmospheric pressure and temperature does not change.

We list the variables:

$V_1 = 2 \text{ L}$ $\qquad\qquad$ $V_2 = ?$

$n_1 = 0.2 \text{ mole}$ $\qquad\qquad$ $n_2 = 0.5 \text{ mole}$

Since the number of moles is increasing, the volume must also increase. We solve for V_2:

$$\frac{V_1}{n_1} = \frac{V_2}{n_2}$$

$$V_2 = V_1 \times \frac{n_2}{n_1}$$

$$V_2 = 2\,L \times \frac{0.5\,\text{mol}}{0.2\,\text{mol}} = 5\,L$$

Next time you blow up a balloon, think of Avogadro.

What is STP?

As you know the pressure and temperature where we live is constantly changing. So if two scientists collected a volume of gas from a chemical reaction at different temperatures and pressures, it would be hard to compare the data. To make it easier to compare data, scientists use standard conditions of temperature and pressure (called STP), which are 0°C (273 K) and 1 atm (760 mmHg).

Molar Volume

From Avogadro's law we know that 1 mole of any gas at the same temperature and pressure will occupy the same volume. At STP (1 atm and 0°C) 1 mole of a gas occupies 22.4 liters. This is called the molar volume of a gas. It does not matter what the gas is. A mole of helium, a mole of nitrogen, or a mole of carbon dioxide, for instance, all have a volume of 22.4 L at STP conditions.

So now we have a new unit conversion for moles:

1 mole of gas at STP = 22.4 L

What is the volume of 200.0 grams of carbon dioxide (CO_2) gas at STP?

The molar mass of CO_2 is 44.0 g/mol. The molar volume of CO_2 is 22.4L at STP.

$$200.0\,\text{g}\,CO_2 \times \frac{1\,\text{mol}\,CO_2}{44.0\,\text{g}\,CO_2} \times \frac{22.4\,L}{1\,\text{mol}\,CO_2} = 102\,L$$

The Ideal Gas Law

The four properties that we use to describe a gas are all related in one equation called the ideal gas law. This equation is derived by combining the combined gas law with Avogadro's law.

$$PV = nRT$$

We can find out the value of R, called the universal gas constant, by solving for R and putting the STP conditions into the equation (1.0 mole of gas with a volume of 22.4 liter at 1 atm and a temperature of 273 K).

$$R = \frac{PV}{nT}$$

$$R = \frac{1\,atm \times 22.4\,L}{1\,mol \times 273\,K} = 0.0821\,atm\,L/mol\,K$$

When we use the ideal gas law, we have to make sure that our units of pressure and volume match those in R. If we know any three of the four variables, we can find the fourth (assuming we know the value of R).

What is the volume of 2.50 moles of a gas at a pressure of 1.50 atm and a temperature of 24°C?

First we make a list of the variables:

$$P = 1.50\,atm \qquad\qquad V = ?$$

$$T = 297\,K\,(24 + 273) \qquad\qquad n = 2.50\,mol$$

Then solve for V.

$$V = \frac{nRT}{P}$$

Now plug in the values for the variables and the constant R:

$$V = \frac{2.5\,mol \times 0.0821\,atm\,L/K\,mol \times 297\,K}{1.50\,atm} = 40.6\,L$$

Now try the practice problems on page 185.

CHAPTER 15

How Strong is That Coffee?
Solution Concentrations

A solution is a uniform mixture of two or more substances. The substance present in the smaller quantity is called the solute and that in the greater quantity, the solvent. Whenever we prepare a solution, we can describe it qualitatively as either concentrated or dilute. A concentrated solution contains more solute in a given amount of solution than does a dilute solution. Strong coffee is a concentrated solution of molecules extracted from coffee beans (the solute molecules) in water (the solvent). Weak "dishwater" coffee contains fewer of the molecules from the coffee beans in a given amount of solution, so it is less concentrated or more dilute than strong coffee.

"Strong" and "weak" are fine descriptions for coffee, but in chemistry we need more precise, quantitative measurements. If you have your blood tested for sodium or potassium, you want a quantity, not just a description of "high" or "low."

The value of the "concentration" of a solution tells us how much solute is dissolved in a particular amount of solution. Note that we do not usually specify the amount of solvent since enough is added to bring the final quantity of solution to the desired value. We can write a word equation as follows:

$$\text{Concentration} = \frac{\text{Quantity of solute}}{\text{Quantity of solution}}$$

In this chapter we will look at the two most common ways to describe concentration: percent and molarity.

Percent Concentrations

Remember that percent is defined as

$$\text{Percent} = \frac{\text{Part}}{\text{Total}} \times 100\%$$

There are three types of percent concentrations for solutions. Sometimes this can be confusing, even for chemists. The three types of percent concentrations differ only in the units used for the total solution. Let's look at each type.

Weight/Volume Percent

In chemistry, as well as in medicine, the weight/volume percent is the percent concentration most commonly used, especially when dissolving solids in liquids. The weight/volume percent is defined as the number of grams of solute dissolved in a given number of milliliters of total solution.

$$\text{Percent (w/v)} = \frac{\text{grams of solute}}{\text{mL of solution}} \times 100\%$$

Calculate the percent (w/v) concentration of a solution in which you've dissolved 4.5 grams of salt, NaCl, into 500 mL of total solution.

Using the formula for weight/volume percent, we calculate

$$\text{Percent (w/v)} = \frac{4.5 \text{ g NaCl}}{500 \text{ mL}} \times 100\% = 0.90\% \text{ (w/v)}$$

This concentration of NaCl is called normal saline. Because it is close to the salt concentration of normal human blood, pharmaceutical solutions that are injected into veins usually are this concentration of NaCl.

The percent (w/v) tell you that every 100 mL of solution will have 0.9 grams of NaCl.

$$0.90\%(\text{w/v}) \quad \text{means} \quad \frac{0.90 \text{ g}}{100 \text{ ml}}$$

Remember "percent" means "out of 100." In weight/volume percents, the 100 is 100 mL.

We can write percent concentrations as conversion factors to solve solution problems. We can write the 0.90% (w/v) we just calculated as the following conversion factors:

$$\frac{0.90 \text{ g NaCl}}{100 \text{ mL}} \quad \text{or} \quad \frac{100 \text{ mL}}{0.90 \text{ g NaCl}}$$

If we are given a volume of solution and want to find how many grams it contains, we use the first conversion. If we must have a certain number of grams and want to find how many milliliters we need, we use the second. Let's do an example of each problem.

A pharmacy technician is preparing 800 mL of 5% (w/v) glucose solution. How many grams of glucose does she need to weigh?

Using the 5% glucose (w/v) as a conversion factor, we calculate:

$$800 \text{ mL} \times \frac{5.0 \text{ g glucose}}{100 \text{ mL}} = 40 \text{ g glucose}$$

In hospitals, this type of solution is called "D5W." D5W strands for 5% dextrose (an alternate form of glucose) in water.

A patient needs to receive 2.5 grams of NaCl in a normal saline solution that is 0.90% (w/v). How many mL of solution will he need?

Since we are given grams, we write the conversion so that the grams cancel and we wind up with mL.

$$2.5 \text{ g NaCl} \times \frac{100 \text{ mL}}{0.90 \text{ g NaCl}} = 278 \text{ mL}$$

This problem shows that if you want to have 2.5 g NaCl, you need to measure out 278 mL of 0.90% (w/v) solution.

Volume/Volume Percents

Since it is easy to measure the volume of liquids, when a liquid is dissolved in another liquid, its solution concentration is usually given as a volume/volume percent. A volume percent is the volume of solute per total volume of solution.

$$\text{Percent (v/v)} = \frac{\text{mL of solute}}{\text{mL of solution}} \times 100\%$$

The hydrogen peroxide solution in your medicine cabinet is a 3% (v/v) solution. Every 100 mL of solution contains 3 mL of hydrogen peroxide. The alcohol content of wine and other alcoholic beverages is reported in volume/volume percents. If a wine label states that it is 15% alcohol, it means every 100 mL of wine contains 15 mL of alcohol. The "proof" of a liquor is twice the volume/volume percent of the alcohol. A 100 proof vodka is 50% (v/v) alcohol.

How much alcohol does someone ingest if he drinks three 30 mL shots of 100 proof (50% (v/v) liquor?

We use the percent of alcohol as a conversion factor:

$$90 \text{ mL liquor} \times \frac{50 \text{ mL alcohol}}{100 \text{ mL liquor}} = 45 \text{ mL alcohol}$$

He is consuming 45 mL of alcohol.

How much beer would a person have to drink to ingest the same amount of alcohol? Most beer is 8% (v/v) alcohol.

In the last problem, we were given a volume of solution (the liquor), and found the amount of solute (the alcohol). In this problem we are given the volume of solute (the alcohol) and asked to find the amount of solution (the beer).

$$45 \text{ mL alcohol} \times \frac{100 \text{ mL beer}}{8 \text{ mL alcohol}} = 563 \text{ mL beer}$$

That is about 1½ twelve ounce cans of beer.

Weight/Weight Percents

The third type of percent concentration is percent (weight/weight). This type of percent is often used in metal alloys, which are mixtures of metals.

$$\text{Percent (w/w)} = \frac{\text{grams of solute}}{\text{grams of total solution}} \times 100\%$$

If a gold alloy is 76% (w/w) gold, it means that every 100 grams of alloy has 76 grams of gold. We can write:

$$76\% \text{ (w/w) gold} = \frac{76 \text{ g gold}}{100 \text{ g alloy}}$$

If you mix 140.0 grams of gold and 200.0 grams of silver, what is the percent (w/w) of gold?

The total solution is 140.0 g gold + 200.0 g silver = 340.0 grams total.

$$\%(w/w) \text{ gold} = \frac{140.0 \text{g gold}}{340.0 \text{ g total}} \times 100 = 41.18\% \text{ gold}$$

We rounded our answer off to 4 significant figures.

If a solution is 5.0% (w/w), how many grams of solute are there in 550.0 grams of solution?

We use the percent (w/w) as a unit conversion.

$$550.0 \text{ g solution} \times \frac{5 \text{ g solute}}{100 \text{ g solution}} = 27.5 \text{ gram solute}$$

Moles Again—Molarity

Molarity is probably the most commonly used concentration in chemistry. Molarity (M) is defined as the moles of solute per one liter of solution. Note: M is not really a unit but an abbreviation for mol/L. If you see a concentration on a label written as 1.5 M, remember that this really is 1.5 mol/L. For example, a 1.0 M solution has 1.0 moles in every liter of solution, a 2.0 M solution has 2.0 moles in every liter of solution, a 0.5 M solution has 0.5 moles of solute in every liter of solution.

$$\text{Molarity}\left(\frac{\text{mol}}{\text{L}}\right) = \frac{\text{amount of solute (mol)}}{\text{volume of solution (L)}}$$

If we dissolve 0.10 moles of HCl in 250 mL of total solution, what is the molarity of the solution?

Since molarity is moles per liter, we will convert the milliliters to liters.

$$250 \text{ mL} \times \frac{1 \text{ L}}{1000 \text{ mL}} = 0.25 \text{ L}$$

Now, dividing the moles of solute by the liters of solution, we calculate:

$$\text{Molarity} = \frac{0.10 \text{ mole HCl}}{0.25 \text{ L}} = \frac{0.40 \text{ mol HCl}}{1 \text{ L}} = 0.40 \text{ M}$$

Conversions Using Molarity

We can use molarity as a unit conversion. Just as we saw with the percent concentrations, there are two types of problems. If we are given a volume of solution, we can find the moles. Or, if we are given moles, we can find the volume of solution.

When we are given a molarity we can write two unit conversions. For example, we can write two unit conversion factors for a 12.0 M HCl:

$$\frac{12.0 \text{ mol HCl}}{1 \text{ L solution}} \quad \text{and} \quad \frac{1 \text{ L solution}}{12.0 \text{ mol HCl}}$$

How many moles of HCl are there in 250 mL of 12.0 M HCl?

We use the molarity as a conversion factor to calculate the moles of HCl after converting the milliliters to liters.

$$250 \text{ mL} \times \frac{1 \text{ L}}{1000 \text{ mL}} \times \frac{12.0 \text{ mol}}{1 \text{ L}} = 3.0 \text{ moles HCl}$$

How many grams of Na_2SO_4 (molar mass = 142 grams) are there in 250 mL of 0.20 M solution?

Again in this problem we are given a volume of solution. We just have one additional step of converting the moles of Na_2SO_4 to grams.

$$250 \text{ mL} \times \frac{1 \text{ L}}{1000 \text{ mL}} \times \frac{0.20 \text{ mol } Na_2SO_4}{1 \text{ L}} \times \frac{142 \text{ g } Na_2SO_4}{1 \text{ mol } Na_2SO_4} = 7.1 \text{ g } Na_2SO_4$$

This problem could have been worded: how many grams of Na_2SO_4 do you need to pre-pare 250 mL of a 0.20 M solution? In either case, we are given a volume of solution, use molarity to find the amount of solute.

Let's try another similar problem.

How many grams of glucose $(C_6H_{12}O_6)$ do you dissolve into 100.0 mL to make a 0.10 M solution? The molar mass of glucose is 180 gram/mol.

$$100.0 \text{ mL} \times \frac{1 \text{ L}}{1000 \text{ mL}} \times \frac{0.10 \text{ mol } C_6H_{12}O_6}{1 \text{ L}} \times \frac{180.0 \text{ g } C_6H_{12}O_6}{1 \text{ mol } C_6H_{12}O_6} = 1.8 \text{ g } C_6H_{12}O_6$$

First we converted ml to L, then we used the molarity as a conversion to go to number of moles of glucose. Finally we used the molar mass of glucose to convert to grams.

We can go in the reverse direction. If we are given a number of moles or grams, we can find the volume of solution of a given molarity in liters or milliliters.

What volume, in liters, of 0.25 M K_2SO_4 (Molar mass = 174.3 grams) solution is needed to provide 50.0 grams of K_2SO_4?

The given quantity is grams and we want to find volume in milliliters.

$$50.0 \text{ g } K_2SO_4 \times \frac{1 \text{ mol } K_2SO_4}{174.2 \text{ g } K_2SO_4} \times \frac{1 \text{ L}}{0.25 \text{ mol } K_2SO_4} = 1.15 \text{ L}$$

If you need 2.0 moles of HCl for a chemical reaction, how many mL of 12.0 M HCl solution do you need?

Write down the given quantity, 2.0 mol HCl, then write the molarity conversion so that the mole units cancel.

$$2.0 \text{ mol HCl} \times \frac{1 \text{ L}}{12.0 \text{ mol HCl}} \times \frac{1000 \text{ mL}}{1 \text{ L}} = 167 \text{ mL}$$

Looking Ahead

Concentrations in molarity are often written with square brackets [ ]. This is especially true in acid/base chemistry, which we will look at in the next chapter. When you see [H^+], it stands for the concentration of H^+ in moles per liter. [OH^-] means the concentration of OH^- in moles per liter. An example is [H^+] = 1.0×10^{-7} mol/L

Now try the practice problems on page 187.

CHAPTER 16

The Basics on a Sour Subject Logarithms and pH

In the last chapter we studied solution concentrations, particularly molarity, which is defined as the number of moles per liter. This chapter looks at concentrations in moles per liter of two of the most important ions in chemistry (and also in our body fluids): the hydrogen ion, written as either H^+ or H_3O^+, and the hydroxide ion, OH^-.

A Little Background on Acids and Bases

Some compounds ionize to produce hydrogen ions (H^+) when added to water. These compounds are called acids. Among the most common acids are H_2SO_4 (sulfuric acid), HNO_3 (nitric acid), HCl (hydrochloric acid), and H_3PO_4 (phosphoric acid). Adding any of these acids to water increases the number of hydrogen ions in the solution.

Nitric acid, for example, ionizes in water as:

$$HNO_3 \rightarrow H^+ + NO_3^-$$

Sulfuric acid ionizes in water as:

$$H_2SO_4 \rightarrow 2H^+ + SO_4^{2-}$$

Compounds such as NaOH (sodium hydroxide), KOH (potassium hydroxide), and $Ca(OH)_2$ (calcium hydroxide) increase the number of hydroxide ions in water solution. These compounds are called bases. When you put sodium hydroxide into water, it ionizes like this:

$$NaOH \rightarrow Na^+ + OH^-$$

Your textbook will go into much more detail about the properties, definitions, and reactions of acids and bases.

When an acid such as HCl ionizes in water, it splits into a hydrogen ion and a chloride ion.

$$HCl \rightarrow H^+ + Cl^-$$

The hydrogen ion, which is nothing but a bare proton, combines with a water molecule to yield H_3O^+, called the hydronium ion, and a chloride ion.

$$HCl + H_2O \rightarrow H_3O^+ + Cl^-$$

So there are two different ways chemists write acids: as H^+, the hydrogen ion, or as H_3O^+, the hydronium ion. In your textbook and in your lectures you will see these two forms used interchangeably.

The Ionization of Water

In pure water a small amount of water molecules will dissociate into ions.

$$H_2O \rightarrow H^+ + OH^-$$

or, alternately we can show the hydrogen ion reacting with another water molecule,

$$H_2O + H_2O \rightarrow H_3O^+ + OH^-$$

Chemists have found that the concentrations of hydrogen ions and hydroxide ions are related by the following equation

$$K_w = [H^+][OH^-]$$

or, using the hydronium ion form,

$$K_w = [H_3O^+][OH^-]$$

This equation states that the product of the concentration of hydronium ions $[H_3O^+]$ and hydroxide ions $[OH^-]$ is a constant whose value is K_w, called the ion product constant. Remember when you see a formula in square brackets [] that it stands for the concentration of what is in the brackets in moles per liter. In pure water, at 25°C, $[H_3O^+]$ and $[OH^-]$ are both equal to 1.0×10^{-7} mol/L. So we can calculate the value of K_w:

$$K_w = (1.0\times10^{-7})(1.0\times10^{-7}) = 1.0\times10^{-14}$$

This is an inverse relationship similar to those in Chapter 7. When $[H_3O^+]$ goes up, $[OH^-]$ goes down; when $[H_3O^+]$ goes down, $[OH^-]$ goes up. Most often you will find that $[H_3O^+]$ and $[OH^-]$ can vary between 1.0×10^0 and 1.0×10^{-14} mol/L.

In problems you will be given the concentration of one ion and asked to find the concentration of the other. If you are given $[H_3O^+]$, you can find $[OH^-]$; if you are given $[OH^-]$, you can find $[H_3O^+]$. You will probably be expected to memorize the value of K_w, 1.0×10^{-14}.

The concentration of $[H_3O^+]$ in a cup of coffee is around 1.0×10^{-5} M. What is $[OH^-]$?

We solve for $[OH^-]$

$$[OH^-] = \frac{K_w}{[H_3O^+]}$$

Plugging in the values for K_w and $[H_3O^+]$:

$$[OH^-] = \frac{1 \times 10^{-14}}{1 \times 10^{-5}} = 1.0 \times 10^{-9}$$

On the calculator we press

1 (EXP) 14 (+/-) (÷) 1 (EXP) 5 (+/-) (=)   The display reads $1.0000000 \; ^{-09}$

This example shows that when the concentration of acid increases from 1.0×10^{-7} in pure water to 1.0×10^{-5} mol/L in the coffee, the concentration of base decreases from 1.0×10^{-7} to 1.0×10^{-9} mol/L. In an inverse relationship, when one number goes up, the other number goes down.

Remember that the larger the negative exponent, the smaller the number actually is. So 10^{-5} is larger than 10^{-7} which is larger than 10^{-9}.

Important
Point

The $[OH^-]$ in a household ammonia solution is 1.9×10^{-5} mol/L. What is $[H_3O^+]$?

We solve for $[H_3O^+]$

$$[H_3O^+] = \frac{K_w}{[OH^-]}$$

Plugging in the numbers for K_w and $[H_3O^+]$

$$[H_3O^+] = \frac{1.0 \times 10^{-14}}{1.9 \times 10^{-5}} = 5.3 \times 10^{-10}$$

On the calculator we press

1 (EXP) 14 (+/-) (÷) 1.9 (EXP) 5 (+/-) (=)   The display reads $5.263157 \; ^{-10}$

We round to two significant figures to get 5.3×10^{-10} mol/L.

This example shows that increasing $[OH^-]$ to 1.9×10^{-5} mol/L caused $[H_3O^+]$ to decrease to 5.3×10^{-10} mol/L. Always, in an inverse relationship, when one value increases, the other value must decrease.

The Advantage of Simple Numbers
What is pH?

As we have seen, the concentration of H_3O^+ can vary in a water solution from the very acidic, 1.0×10^0 mol/L, to the very basic, 1×10^{-14} mol/L, with pure water in the middle at 1.0×10^{-7} mol/L. It can be awkward to say the concentration of acid in moles per liter. So chemists invented a short cut to state $[H_3O^+]$. It is called pH, which is defined as the negative of the logarithm of the H_3O^+ concentration in moles per liter.

$$pH = -\log[H_3O^+]$$

The concentration of H_3O^+ in normal human blood is 4.0×10^{-8} mol/L. If someone's concentration increases to just 5.5×10^{-8} mol/L, he can be in serious trouble. It is much easier to say "the patient's pH is 7.40."or "the patient's pH is 7.26." So pH is clearly a convenient short cut.

Solutions with pH values below 7 are acidic. The lower the pH, the more concentrated the acid. Milk, with a pH of about 6, is slightly acidic. Lemon juice, with a pH of 2, is much more acidic. Solutions with pH values greater than 7 are basic. The higher the pH the more basic the solution. Blood, with a pH of 7.4, is slightly basic. An ammonia solution, with a pH of 11, is much more basic. Before we look at calculations involving pH, let's look at what logarithms are and how you handle them on a calculator.

Just What is a Logarithm Anyway?

"Logarithm"is just another word for "exponent." If you write 10^{-3}, you already know that the base is 10 and the exponent is -3. You can also refer to the -3 as the logarithm of 10^{-3} or, stating it more simply, as the log of 10^{-3}. You can write this expression as

$$\log(10^{-3}) = -3$$

Now remember that you can write 10^{-3} with a coefficient of 1:

$$10^{-3} = 1 \times 10^{-3}$$

Therefore if you were asked to find the logarithm of 10^{-3}, the problem could be written as

$$\log(10^{-3}) = -3 \quad \text{or} \quad \log(1\times10^{-3}) = -3$$

What is the logarithm of 10^{-7}?

$$\log(10^{-7}) = -7 \quad \text{or} \quad \log(1\times10^{-7}) = -7$$

What is the logarithm of 10^{-13}?

$$\log(10^{-13}) = -13 \quad \text{or} \quad \log(1\times10^{-13}) = -13$$

What is the logarithm of 10^{0}? Remember $10^{0} = 1$, so we can write this problem as

$$\log(10^{0}) = 0 \quad \text{or} \quad \log(1\times10^{0}) = 0 \quad \text{or} \quad \log(1) = 0$$

So for any power of 10, where the coefficient is 1, the logarithm is simply the whole number exponent. If we put a coefficient other than 1 in front of the power of 10, the problem is not quite so simple. But here is where your calculator comes in handy. Let's calculate

$$\log(5.0\times10^{-7})$$

First enter the number in scientific notation.

5 (EXP) 7 (+/-)   The display reads   5. $^{-07}$

Now press (log). This button calculates the logarithm for you.

The display now reads -6.3010

We round off to get -6.30.

Important Point

Note on significant figures in logarithms. The logarithm of a number is reported with the same number of significant figures *after the decimal point* as there are in the whole number. In the example above, 5.0×10^{-7} has two significant figures, so we rounded the answer to two significant figures after the decimal point to -6.30.

Warning: your calculator also has a (LN) button. This button takes "natural logarithms" that use a base other than 10. You will always use the (log) button, never the (LN) button.

What is the logarithm of 6.50×10^{-11}?

6.5 (EXP) 11 (+/-)   The display reads   6.5 $^{-11}$

Press $\boxed{\text{log}}$ The display reads -10.18708

We round to three significant figures past the decimal point to get -10.187.
This answer means $6.50 \times 10^{-11} = 10^{-10.187}$

In chemistry you will use the logarithms that have a base of 10. However, the base of a logarithm does not have to be 10. The base can be any other number. For example $2^3 = 8$. We can write this in logarithm form as $\log_2 (8) = 3$. The base is now 2, so it is written as a subscript under the log sign. If there is no subscript shown, the base is assumed to be 10.

Going in Reverse

Now suppose you are given the logarithm and are asked to find its value. In other words we want to "undo" the logarithm. Suppose the logarithm of a number is -4.20. What is the number?

If the logarithm of a number is -4.20, then the number written as a power of 10 is $10^{-4.20}$.

So the problem is really asking, "What number is equal to $10^{-4.20}$?" Above the $\boxed{\text{log}}$ button on your calculator you will see 10^x. This function calculates the number if you enter its logarithm. You get to this function by pressing the $\boxed{\text{INV}}$ button in the upper left hand corner. On other calculators the button may be $\boxed{\text{2nd}}$ or $\boxed{\text{SHIFT}}$. When you press $\boxed{\text{INV}}$ $\boxed{\text{log}}$ you get the 10^x function written above the $\boxed{\text{log}}$ button. This "undoes" a logarithm.

So on the calculator we press

$4.2 \boxed{+/-}$ The display reads -4.2

Then press $\boxed{\text{INV}}$ $\boxed{\text{log}}$ The display reads $6.30957 \ ^{-05}$

We round the answer to 2 significant figures and get 6.3×10^{-5}.

The logarithm of a number is -9.20. What is the number?

If the logarithm is -9.20, this really says the number will be $10^{-9.20}$.

We enter 9.2 $\boxed{+/-}$ then press $\boxed{\text{INV}}$ $\boxed{\text{log}}$.   The display reads $6.30957 \ ^{-10}$

We round the answer to 2 significant figures and get 6.3×10^{-10}.

Warning: do not try to use the $\boxed{\text{EXP}}$ key to "undo" the logarithm. This key simply multiplies what you have entered by 10 to a whole number power. It will not work with decimal exponents.

The logarithm of a number is –5.30. What is the number?

The number will be equal to $10^{-5.30}$.

Press 5.30 $\boxed{+/-}$ then press $\boxed{\text{INV}}$ $\boxed{\text{log}}$. The display reads 5.01187^{-06}

We round our answer off to $5.0{\times}10^{-6}$.

The logarithm of a number is –6. What is the number?

Since the exponent is a whole number, we don't need a calculator to do this problem. The number is simply 10^{-6} or $1{\times}10^{-6}$. But if we do plug it in the calculator:

Press 6 $\boxed{+/-}$ then press $\boxed{\text{INV}}$ $\boxed{\text{log}}$. The display reads 1.0000000^{-06}

Back to pH

Now that we've taken a look at what logarithms are, let's calculate the pH values of some solutions.

First let's look at the pH values of whole powers of 10. We can do these without a calculator. In the next section we will look at problems that will need a calculator.

The $[H_3O^+]$ of a solution is 10^{-3} mol/L. What is the pH?

$$pH = -\log[H_3O^+] = -\log(10^{-3}) = -(-3) = 3$$

The negative sign in front of the log simply turns the pH into a positive value. (Remember that two negative numbers multiplied together make a positive.) It is easier to say "the pH is 3" than it is to say "the pH is negative 3." This particular solution is acidic.

What is the pH of a solution where $[H_3O^+] = 10^{-11}$ mol/L?

$$pH = -\log(10^{-11}) = -(-11) = 11$$

The logarithm of 10^{-11} is simply –11. This solution is basic.

In pure water the $[H_3O^+]$ is $1.0{\times}10^{-7}$ mol/L? What is the pH?

$$pH = -\log(10^{-7}) = -(-7) = 7$$

A solution with the pH at exactly 7 is neutral.

Back in Reverse

Now let's go the other way. The pH is 8; what is the H_3O^+ concentration?

$$[H_3O^+] = 10^{-pH} = 10^{-8} \text{ mol/L or } 1.0 \times 10^{-8} \text{ mol/L}$$

Remember pH is a unitless number. When you convert to concentration, you have to tag on the units of moles per liter.

The pH is 1. What is $[H_3O^+]$?

$$[H_3O^+] = 10^{-pH} = 10^{-1} \text{ mol/L or } 1.0 \times 10^{-1} \text{ mol/L}$$

Because this number is an ordinary size number, it can also be written as in decimal form as 0.1 mol/L.

The pH of a solution is 0. What is $[H_3O^+]$?

$$[H_3O^+] = 10^{-pH} = 10^{0} \text{ mol/L or } 1 \text{ mol/L}$$

Remember $10^{0} = 1$. If a solution has a pH of 0, the concentration of acid is 1 mol/L. This solution is very acidic.

Thank Goodness for My Calculator
pH Calculations

Now we will try a few problems that involve coefficients other than 1.

The $[H_3O^+]$ of a solution is 5.0×10^{-9} mol/L. What is the pH?

$$pH = -\log(5.0 \times 10^{-9})$$

Enter: 5 (EXP) 9 (+/-) then enter (log) (+/-)   The display reads   *8.30103*

We round to 2 significant figures after the decimal point to get 8.30.

The second time we press (+/-) after we press (log) simply changes negative 8.3 to positive 8.3. This solution is slightly basic.

Blood, as we said earlier, has a $[H_3O^+]$ of 4.0×10^{-8} mol/L. What is the pH?

$$pH = -\log(4.0 \times 10^{-8})$$

Enter 4 $\boxed{\text{EXP}}$ 8 $\boxed{\text{+/-}}$ then enter $\boxed{\text{log}}$ $\boxed{\text{+/-}}$ The display reads 7.39794

We round to 2 significant figures after the decimal point to get 7.40.

The H_3O^+ concentration is 7.49×10^{-11} mol/L. What is the pH?

$$pH = \log(7.49 \times 10^{-11})$$

Enter 7.49 $\boxed{\text{EXP}}$ 11 $\boxed{\text{+/-}}$ then enter $\boxed{\text{log}}$ $\boxed{\text{+/-}}$ The display reads 10.1255

We round our answer to 3 significant figures after the decimal point to 10.126.

Back in Reverse

Now we will go the other way. In these problems you are given the pH and asked to find the H_3O^+ concentration.

The pH of orange juice is 3.10. What is the H_3O^+ concentration?

$$[H_3O^+] = 10^{-pH}$$

Enter 3.1, then press $\boxed{\text{+/-}}$ to make it negative.

Now press $\boxed{\text{INV}}$ $\boxed{\text{log}}$ to get to the second function 10^x. This "undoes" the logarithm.

The display reads 7.94328 $^{-04}$

We round off to 2 significant figures to get 7.9×10^{-4} mol/L.

Acid rain can have pH values lower than 4.5. What is the H_3O^+ concentration?

$$[H_3O^+] = 10^{-pH} = 10^{-4.5}$$

Enter 4.5, press $\boxed{\text{+/-}}$ to make it negative.

Now press $\boxed{\text{INV}}$ $\boxed{\text{log}}$ to get to the second function 10^x. The display reads 3.162277 $^{-05}$

We round our answer to 1 significant figure to get 3×10^{-5} mol/L.

Let's try another.

The pH of a solution is 8.92. Find $[H_3O^+]$ and then calculate $[OH^-]$.

This problem has two parts. First we calculate $[H_3O^+]$.

$$[H_3O^+] = 10^{-pH}$$

Enter 8.92 $\boxed{+/-}$ then press $\boxed{INV}$ $\boxed{log}$ The display reads 1.202264 $^{-09}$

We round the answer to 2 significant figures to get $[H_3O^+] = 1.2 \times 10^{-9}$ mol/L.

Next write the equation:

$$K_w = [H_3O^+][OH^-]$$

Now solve for $[OH^-]$:

$$[OH^-] = \frac{K_w}{[H_3O^+]}$$

Then plug in K_w and the value for $[H_3O^+]$ that we just calculated:

$$[OH^-] = \frac{1.00 \times 10^{-14}}{1.20 \times 10^{-9}} = 8.33 \times 10^{-6}$$

Notice the relationship between $[H_3O^+]$ and $[OH^-]$: $[H_3O^+]$ is less than 1.0×10^{-7} mol/L at 1.20×10^{-9} mol/L and so $[OH^-]$ has to be greater than 1.0×10^{-7} at 8.3×10^{-6} mol/L.

Now try the practice problems on page 189.

Chapter 2 Practice Problems

1. $8 + (-4) =$

2. $8 - (-4) =$

3. $8(-4) =$

4. $\dfrac{8}{-4} =$

5. $-8 + (-4) =$

6. $-8 - (-4) =$

7. $-8(-4) =$

8. $\dfrac{-4}{-8} =$

9. $\dfrac{1}{2}$ of $\dfrac{1}{3} =$

10. $\dfrac{1}{2} \div \dfrac{1}{3} =$

11. $\dfrac{1}{2}$ of $8 =$

12. $\dfrac{1}{2} \times 8 =$

13. $\dfrac{1}{2} \div 8 =$

14. $8 \div \dfrac{1}{2} =$

15 Convert $\dfrac{1}{2}$ to a decimal.

16. Convert $\dfrac{11}{5}$ to a decimal.

17. Convert $\dfrac{3}{10}$ to a decimal.

18. Convert $\dfrac{15}{100}$ to a decimal.

19. Convert 13% to a decimal.

20. Convert 1.3% to a decimal.

21. If you get 40 out of 50 problems correct on a test, what is your percent grade?

22. If you want to score 90% on a 60 problem test, how many problems do you need to get correct?

23. $10^5 \times 10^8 =$

24. $\dfrac{10^5}{10^8} =$

25. $10^{-2} \times 10^{-1} =$

26. $\dfrac{10^{-2}}{10^{-1}} =$

27. Convert 8×10^7 to decimal form.

28. Convert 1.3×10^{-3} to decimal form.

29. Convert 289,000 to scientific notation.

30. Convert 0.000054 to scientific notation.

Solutions to Chapter 2 Problems

1. 4

2. $8 - (-4) = 8 + 4 = 12$

2. −32

4. −2

5. −12

6. $-8 - (-4) = -8 + 4 = -4$

7. 32

8. 1/2 or 0.5

9. $\frac{1}{2} \times \frac{1}{3} = \frac{1}{6}$

10. $\frac{1}{2} \div \frac{1}{3} = \frac{1}{2} \times \frac{3}{1} = \frac{3}{2}$ or 1.5

11. $\frac{1}{2}$ of $8 = \frac{1}{2} \times 8 = 4$

12. 4 (11 and 12 are the same problem)

13. $\frac{1}{2} \div 8 = \frac{1}{2} \times \frac{1}{8} = \frac{1}{16}$

14. $8 \div \frac{1}{2} = 8 \times 2 = 16$

15. 0.5

16. 2.2

17. 0.3

18. 0.15

19. 0.13

20. 0.013

21. Percent $= \dfrac{\text{part}}{\text{whole}} \times 100 = \dfrac{40}{50} \times 100 = 80\%$

22. Part $= \dfrac{\%}{100} \times \text{Total} = \dfrac{90}{100} \times 60 = 54$ problems

23. 10^{13}

24. 10^{-3} or $\dfrac{1}{10^3}$ or $\dfrac{1}{1000}$

25. 10^{-3} or $\dfrac{1}{10^3}$ or $\dfrac{1}{1000}$

26. 10^{-1} or $\dfrac{1}{10}$

27. 80,000,000

28. 0.0013

29. 2.89×10^{5}

30. 5.4×10^{-5}

Chapter 3 Practice Problems

Use your calculator. Round all answers to the appropriate number of significant figures.

1. $\dfrac{4.807 \times 1.85}{0.467 \times 5.00} =$

2. $\dfrac{96.8}{15.2 \times 0.0120} =$

3. $10.7 \times \dfrac{1.2}{3.8} \times \dfrac{14.2}{5.01} =$

4. $(4.16 \times 10^5)(5.08 \times 10^{-8}) =$

5. $\dfrac{7.00 \times 10^{-6}}{5.871 \times 10^4} =$

6. $\dfrac{5.68 \times 10^3}{5.02 \times 10^2} =$

7. $(10^5)(10^{-8})(10^3) =$

8. $\dfrac{10^{-14}}{10^{-9}} =$

9. $3.15 \times 10^{18} \times \dfrac{342}{6.02 \times 10^{23}} =$

10. $1.00 \times 10^{15} \times \dfrac{1.66 \times 10^{-24}}{12.0} =$

Solutions to Chapter 3 Problems

1.   4.807 $\boxed{\times}$ 1.85 $\boxed{\div}$ 0.467 $\boxed{+}$ 5 $\boxed{=}$ *3.808543897*

   Rounded to 3 sig figs, the answer is 3.81.

2.   96.8 $\boxed{\div}$ 15.2 $\boxed{\div}$ 0.012 $\boxed{=}$ *530.7017544*

   Rounded to 3 sig figs, the answer is 531.

3.   10.7 $\boxed{\times}$ 1.2 $\boxed{\div}$ 3.8 $\boxed{\times}$ 14.2 $\boxed{\div}$ 5.01 $\boxed{=}$ *9.577056413*

   Rounded to 2 sig figs, the answer is 9.6

4.   4.16 $\boxed{\text{EXP}}$ 5 $\boxed{\times}$ 5.08 $\boxed{\text{EXP}}$ 8 $\boxed{+/-}$ $\boxed{=}$ *0.0211328*

   Rounded to 3 sig figs, the answer is 0.0211. In scientific notation: 2.11×10^{-2}

5.   7 $\boxed{\text{EXP}}$ 6 $\boxed{+/-}$ $\boxed{+}$ 5.871 $\boxed{\text{EXP}}$ 4 $\boxed{=}$ *1.1923011 ⁻¹⁰*

   Rounded to 3 sig figs, the answer is 1.19×10^{-10}

6.   5.68 $\boxed{\text{EXP}}$ 3 $\boxed{\div}$ 5.02 $\boxed{\text{EXP}}$ 2 $\boxed{=}$ *11.31474104*

   Rounded to 3 sig figs, the answer is 11.3. In scientific notation: 1.13×10^{1}

7.   1 $\boxed{\text{EXP}}$ 5 $\boxed{\times}$ 1 $\boxed{\text{EXP}}$ 8 $\boxed{+/-}$ $\boxed{\times}$ 1 $\boxed{\text{EXP}}$ 3 $\boxed{=}$ *1*

   The answer is simply 1. Remember to enter the implied coefficients of 1.

8.   1 $\boxed{\text{EXP}}$ 14 $\boxed{+/-}$ $\boxed{\div}$ 1 $\boxed{\text{EXP}}$ 9 $\boxed{+/-}$ $\boxed{=}$ *0.00001*

   The answer in scientific notation is 1×10^{-5}

9.   3.15 $\boxed{\text{EXP}}$ 18 $\boxed{\times}$ 342 $\boxed{\div}$ 6.02 $\boxed{\text{EXP}}$ 23 $\boxed{=}$ *1.7895349 ⁻⁰³*

   Rounded to 3 sig figs, the answer is 1.79×10^{-3}. The decimal form is 0.00179

10.   1 $\boxed{\text{EXP}}$ 15 $\boxed{\times}$ 1.66 $\boxed{\text{EXP}}$ 24 $\boxed{+/-}$ $\boxed{\div}$ 12 $\boxed{=}$ *1.38333 ⁻¹⁰*

   Rounded to 3 sig figs, the answer is 1.38×10^{-10}

Chapter 5 Practice Problems

1. Convert 4.20 feet to inches.

2. Convert 78 inches to feet.

3. Convert 1.9 meters to centimeters.

4. Convert 526 millimeters to meters.

5. Convert 6.09 kg to grams.

6. Convert 38 milliliters to liters.

7. Convert 0.5 kilometers to centimeters.

8. Convert 7.68×10^5 milligrams to kilograms.

9. Convert 2.50 inches to centimeters.

10. Convert 6.0 fluid ounces to milliliters

Solutions to Chapter 5 Problems

1. $4.2 \, \text{feet} \times \dfrac{12 \, \text{inches}}{1 \, \text{foot}} = 50.4 \, \text{inches}$

2. $78 \, \text{inches} \times \dfrac{1 \, \text{foot}}{12 \, \text{inches}} = 6.5 \, \text{feet}$

3. $1.9 \, \text{m} \times \dfrac{100 \, \text{cm}}{1 \, \text{m}} = 190 \, \text{cm}$

4. $526 \, \text{mm} \times \dfrac{1 \, \text{m}}{1000 \, \text{mm}} = 0.526 \, \text{m}$

5. $6.09 \, \text{kg} \times \dfrac{1000 \, \text{g}}{1 \, \text{kg}} = 6090 \, \text{g}$

6. $38.0 \, \text{mL} \times \dfrac{1 \, \text{L}}{1000 \, \text{mL}} = 0.038 \, \text{L}$

7. $0.5 \, \text{km} \times \dfrac{1000 \, \text{m}}{1 \, \text{km}} \times \dfrac{100 \, \text{cm}}{1 \, \text{m}} = 50{,}000 \, \text{cm} = 5 \times 10^{4} \, \text{cm}$

8. $7.68 \times 10^{5} \, \text{mg} \times \dfrac{1 \, \text{g}}{1000 \, \text{mg}} \times \dfrac{1 \, \text{kg}}{1000 \, \text{g}} = 0.768 \, \text{kg}$

9. $2.50 \, \text{inches} \times \dfrac{2.54 \, \text{cm}}{1 \, \text{inch}} = 6.35 \, \text{cm}$

10. $6.0 \, \text{fl. oz} \times \dfrac{1 \, \text{qt}}{32 \, \text{fl. oz}} \times \dfrac{1 \, \text{L}}{1.06 \, \text{qt}} \times \dfrac{1000 \, \text{mL}}{1 \, \text{L}} = 177 \, \text{mL}$

Chapter 7 Practice Problems

1. Solve $4x - 3 = 25$

2. Solve $7x + 5 = 26$

3. Given $\dfrac{x_1}{y_1} = \dfrac{x_2}{y_2}$, if y changes from 12 to 27 and x is initially 15, what is x_2?

4. Given $x_1 y_1 = x_2 y_2$, if y_1 is 10 and x changes from 3 to 6, how does y change?

5. Given $x_1 y_1 = x_2 y_2$, if y changes from 9 to 38, and x_2 is 15, what was x_1?

6. Given $\dfrac{x_1}{y_1} = \dfrac{x_2}{y_2}$, if x changes from 2 to 8 and y is initially 3, what is the new value of y?

Solutions to Chapter 7 Problems

1. $4x - 3 = 25$

   $4x - 3 + 3 = 25 + 3$

   $4x = 28$

   $x = 7$

2. $7x + 5 = 26$

   $7x + 5 - 5 = 26 - 5$

   $7x = 21$

   $x = 3$

3. $x_1 = 15$ $x_2 = ?$

   $y_1 = 12$ $y_2 = 27$

   $$x_2 = \frac{x_1 \times y_2}{y_1} = \frac{15 \times 27}{12} = 33.75 \quad \text{Rounded to 2 sig figs the answer is 34.}$$

   Notice that in this direct relationship as y increased, x also increased.

4. $x_1 = 3$ $x_2 = 6$

   $y_1 = 10$ $y_2 = ?$

   $$y_2 = \frac{y_1 \times x_1}{x_2} = \frac{10 \times 3}{6} = 5$$

   Notice that in this inverse relationship as x increased, y decreased.

5. $x_1 = ?$ $x_2 = 15$

   $y_1 = 9$ $y_2 = 38$

   $$x_1 = \frac{x_2 \times y_2}{y_1} = \frac{15 \times 38}{9} = 63.3$$

6. $x_1 = 2$ $x_2 = 8$

   $y_1 = 3$ $y_2 = ?$

   Cross multiply to get $x_1 y_2 = x_2 y_1$

   $$y_2 = \frac{x_2 y_1}{x_1} = \frac{8 \times 3}{2} = 12 \quad \text{Notice as } x \text{ increased, } y \text{ also increased.}$$

Chapter 8 Practice Problems

1.   If 20.0 mL of a liquid weighs 17.4 g, what is its density?

2.   Calculate the density of a piece of lead that has a mass of 282.5 grams and occupies a volume of 25.0 cm^3.

3.   Calculate the mass of 0.50 cm^3 of gold. (See Table 8.1 for the density of gold.)

4.   What is the mass of 75.0 cm^3 of aluminum? (See Table 8.1)

5.   What is the mass in kilograms of 2.0 liters of gasoline? (See Table 8.1)

6.   What is the volume of 850.0 g of silver? (See Table 8.1)

7.   Find the volume in liters of 900.0 g of ethyl alcohol. (See Table 8.1)

8.   What volume of mercury is needed to obtain 210.0 grams of the metal? (See Table 8.1)

9.   If 10.0 mL of a chemical has a mass of 15.8 g, what is its specific gravity?

10.   Ice at 4°C has a specific gravity of 0.92. What is the mass of 55.0 cm^3 of ice?

Solutions to Chapter 8 Problems

1. $\dfrac{17.4\,\text{g}}{20.0\,\text{mL}} = 0.870\,\text{g/mL}$

2. $\dfrac{282.5\,\text{g}}{25.0\,\text{cm}^3} = 11.3\,\text{g/cm}^3$

3. $0.50\,\cancel{\text{cm}^3} \times \dfrac{19.6\,\text{g}}{1\,\cancel{\text{cm}^3}} = 9.8\,\text{g}$

4. $75.0\,\cancel{\text{cm}^3} \times \dfrac{2.70\,\text{g}}{1\,\cancel{\text{cm}^3}} = 203\,\text{g}$

5. $2.0\,\cancel{\text{L}} \times \dfrac{1000\,\cancel{\text{mL}}}{1\,\cancel{\text{L}}} \times \dfrac{0.66\,\cancel{\text{g}}}{1\,\cancel{\text{mL}}} \times \dfrac{1\,\text{kg}}{1000\,\cancel{\text{g}}} = 1.3\,\text{kg}$

6. $850.0\,\cancel{\text{g}} \times \dfrac{1\,\text{cm}^3}{10.5\,\cancel{\text{g}}} = 81.0\,\text{cm}^3$

7. $900.0\,\cancel{\text{g}} \times \dfrac{1\,\cancel{\text{mL}}}{0.79\,\cancel{\text{g}}} \times \dfrac{1\,\text{L}}{1000\,\cancel{\text{mL}}} = 1.1\,\text{L}$

8. $210.0\,\cancel{\text{g}} \times \dfrac{1\,\text{mL}}{13.6\,\cancel{\text{g}}} = 15.4\,\text{mL}$

9. $\dfrac{15.8\,\text{g}}{10.0\,\text{mL}} = 1.58\,\text{g/mL}$ The specific gravity is 1.58.

10. Since the specific gravity is 0.92, the density is $0.92\,\text{g/cm}^3$.

    Now use the density as a conversion:

    $55.0\,\cancel{\text{cm}^3} \times \dfrac{0.92\,\text{g}}{1\,\cancel{\text{cm}^3}} = 51\,\text{g}$

Chapter 9 Practice Problems

1. If the temperature is 33 °C, will you need to wear a sweater? What is the temperature in °F?

2. Convert –12 °C to °F.

3. If you have a temperature of 102 °F, what will it be in °C ?

4. Convert 80 °F to °C.

5. If the wind chill factor is –15 °F, what is the temperature in °C ?

6. What is the Kelvin temperature of 40 °C ?

Solutions to Chapter 9 Problems

1. $°F = 1.8°C + 32 = 1.8(33) + 32 = 91°F$

    You probably won't need a sweater!

2. $°F = 1.8°C + 32 = 1.8(-12) + 32 = 10°F$

3. $°C = \dfrac{°F - 32}{1.8} = \dfrac{102 - 32}{1.8} = 38.9°C$

4. $°C = \dfrac{°F - 32}{1.8} = \dfrac{80 - 32}{1.8} = 27°C$

5. $°C = \dfrac{°F - 32}{1.8} = \dfrac{-15 - 32}{1.8} = -26°C$

6. $K = °C + 273 = 40 + 273 = 313K$

Chapter 10 Practice Problems

1. A beaker containing 325 g of water is heated from 20 °C to 100 °C. How many calories of heat are required? Find the specific heat of water in Table 10.1.

2. Calculate the heat required to raise the temperature of 44.9 grams of copper from 22.4 °C to 88.9 °C. Find the specific heat of copper in Table 10.1.

3. Calculate the amount of heat released when 98.1 g of silver is cooled from 221.0 °C to 144.0 °C. Find the specific heat of silver in Table 10.1.

4. How much heat is required to heat 500 g of water (about a pint) from 20 °C to 100 °C? Look up the specific heat of water in Table 10.1

5. How much heat is required to heat an aluminum cooking pan that weighs 500 g from room temperature (20 °C) to the boiling point of water (100 °C)? Look up the specific heat of aluminum in Table 10.1.

6. Which required more heat in the two problems above? Why?

7. A 55.42 g sample of a substance absorbed 282 calories when it was heated from 20.0 °C to 72.4 °C. What is the specific heat of the substance?

8. A piece of glass weighing 23.21 grams absorbs 82.02 calories when it is heated from 19.0 °C to 38.0 °C. What is the specific heat of the glass?

9. A 15.0 g sample of one of the substances in Table 10.1 absorbs 218 calories when it is raised in temperature from 22 °C to 88 °C. What is its specific heat and which substance is it?

Solutions to Chapter 10 Problems

1. $\text{Heat} = \text{mass(g)} \times \Delta T(°C) \times \text{specific heat}\left(\frac{\text{cal}}{\text{g°C}}\right)$

    $\text{Heat} = 325\,\text{g} \times 80\,°C \times \dfrac{1.00\,\text{cal}}{\text{g°C}} = 26{,}000\,\text{cal} = 2.6 \times 10^4\,\text{cal}$

2. $\text{Heat} = 44.9\,\text{g} \times 66.5\,°C \times \dfrac{0.092\,\text{cal}}{\text{g°C}} = 280\,\text{cal} = 2.80 \times 10^2\,\text{cal}$

3. $\text{Heat} = 98.1\,\text{g} \times 77\,°C \times \dfrac{0.057\,\text{cal}}{\text{g°C}} = 430\,\text{cal} = 4.3 \times 10^2\,\text{cal}$

4. $\text{Heat} = 500\,\text{g} \times 80\,°C \times \dfrac{1.0\,\text{cal}}{\text{g°C}} = 40{,}000\,\text{cal} = 4.0 \times 10^4\,\text{cal}$

5. $\text{Heat} = 500\,\text{g} \times 80\,°C \times \dfrac{0.22\,\text{cal}}{\text{g°C}} = 8800\,\text{cal} = 8.8 \times 10^3\,\text{cal}$

6. The water took almost five times as many calories to increase the same temperature as the aluminum because its higher specific heat is almost 5 times higher.

7. $\text{Specific heat} = \dfrac{\text{Heat}}{\text{mass} \times \Delta T}$

    $\text{Specific heat} = \dfrac{282\,\text{cal}}{55.42\,\text{g} \times 52.4\,°C} = 0.0971\,\text{cal}/\text{g°C}$

8. $\text{Specific heat} = \dfrac{82.02\,\text{cal}}{23.21\,\text{g} \times 19\,°C} = 0.186\,\text{cal}/\text{g°C}$

9. $\text{Specific heat} = \dfrac{218\,\text{cal}}{15.0\,\text{g} \times 66\,°C} = 0.22\,\text{cal}/\text{g°C}$

    The substance is aluminum.

Chapter 12 Practice Problems

You will need to use your periodic table to calculate the molar mass in some of the problems.

1. What is the molar mass of phenol(C_6H_6O)?

2. What is the molar mass of Iron (III) oxide, Fe_2O_3, the principal component of rust?

3. How many moles are 27.0 g of KF?

4. Aspirin has a molecular formula of $C_9H_8O_4$. What is the mass of 0.220 mole of aspirin?

5. An typical aspirin tablet contains 325 mg of aspirin. How many moles is this?

6. A pint of water (H_2O) weighs about 472 grams. How many moles is this?

7. What is the mass of 5.00 moles of O_2?

8. How many molecules are there in 7.25 moles of CH_4?

9. How many molecules are there in a microscopic speck of carbon soot that weighs one billionth of a gram (1.0×10^{-9} g C)?

Solutions to Chapter 12 Problems

1.
6 mole C atoms	$6 \times 12.0g$ =	72.0g
6 mole H atoms	$6 \times 1.0 g$ =	6.0 g
1 mole O atoms	$1 \times 16.0 g$ =	16.0g
		94.0g

2.
2 mole Fe atoms	$2 \times 55.85g$ =	111.7g
3 mole O atoms	$3 \times 16.0g$ =	48.0g
		159.7 g

3. $27.0 \, g \, \cancel{KF} = \dfrac{1 \, mol \, KF}{58.1 \, g \, \cancel{KF}} = 0.465 \, mol \, KF$

4. The molar mass of aspirin is 180.0 to 4 sig figs.

$$0.220 \, \cancel{mol \, C_9H_8O_4} \times \dfrac{180.0 \, g \, C_9H_9O_4}{1 \, \cancel{mol \, C_9H_9O_4}} = 39.6 \, g \, C_9H_9O_4$$

5. $325 \, \cancel{mg \, C_9H_9O_4} \times \dfrac{1 \, \cancel{g \, C_9H_9O_4}}{1000 \, \cancel{mg \, C_9H_9O_4}} \times \dfrac{1 \, mol \, C_9H_9O_4}{180.0 \, \cancel{g \, C_9H_9O_4}}$

$= 0.00181 \, mol \, C_9H_9O_4$

6. $472 \, \cancel{g \, H_2O} \times \dfrac{1 \, mol \, H_2O}{18.0 \, \cancel{g \, H_2O}} = 26.2 \, mol \, H_2O$

7. $5.00 \, \cancel{mol \, O_2} \times \dfrac{32.0 \, g \, O_2}{1 \, \cancel{mol \, O_2}} = 160 \, g \, O_2$

8. $7.25 \, \cancel{mol \, CH_4} \times \dfrac{6.02 \times 10^{23} \, molecules \, CH_4}{1 \, \cancel{mol \, CH_4}} = 4.36 \times 10^{24} \, molecules \, CH_4$

9. $1.0 \times 10^{-9} \, \cancel{g \, C} \times \dfrac{1 \, \cancel{mol \, C}}{12.0 \, \cancel{g \, C}} \times \dfrac{6.02 \times 10^{23} \, atoms \, C}{1 \, \cancel{mol \, C}} = 5.0 \times 10^{13} \, atoms \, C$

There are 50 trillion atoms in a microscopic speck of carbon.

Chapter 13 Practice Problems

You will need to use your periodic table to calculate the molar mass in some of the problems.

1. How many moles of carbon dioxide are (CO_2) are produced when 2.5 moles of glucose ($C_6H_{12}O_6$) are metabolized in the following reaction?

$$C_6H_{12}O_6 + 6O_2 \rightarrow 6CO_2 + 6H_2O$$

2. When wine ferments, glucose is converted into ethyl alcohol (C_2H_6O) by the following reaction.

$$C_6H_{12}O_6 \rightarrow 2C_2H_6O + 2CO_2$$

   a. When 5.0 moles of glucose are reacted, how many moles of carbon dioxide are produced?

   b. When 250 g of glucose (a little over a half-pound) are fermented, how many grams of ethyl alcohol are produced?

3. Many commercial drain cleaners contain aluminum metal and sodium hydroxide (NaOH). When added to water, the mixture reacts by the following equation to produce hydrogen gas (H_2) and heat that help to dislodge material clogging the drain.

$$2Al + 6NaOH \rightarrow 2Na_3AlO_3 + 3H_2$$

   a. When 2.5 moles of Al react, how many moles of H_2 are produced?

   b. When 56 g of Al react, how many grams of NaOH also react?

Solutions to Chapter 13 Problems

1.
$$C_6H_{12}O_6 + 6O_2 \;\checkmark\; \rightarrow 6CO_2 + 6H_2O$$

$$2.5 \,\text{mol}\,C_6H_{12}O_6 \times \frac{6\,\text{mol CO}_2}{1\,\text{mol}\,C_6H_{12}O_6} = 15\,\text{mol CO}_2$$

2.   a.
$$C_6H_{12}O_6 \;\rightarrow\; 2C_2H_6O + 2CO_2$$

$$5.0\,\text{mol}\,C_6H_{12}O_6 \times \frac{2\,\text{mol CO}_2}{1\,\text{mol}\,C_6H_{12}O_6} = 10\,\text{mol CO}_2$$

   b.
$$C_6H_{12}O_6 \;\rightarrow\; 2C_2H_6O + 2CO_2$$

The molar mass of $C_6H_{12}O_6$ is 180.0 g. The molar mass of C_2H_6O is 46.0 g.

$$250.0\,\text{g}\,C_6H_{12}O_6 \times \frac{1\,\text{mol}\,C_6H_{12}O_6}{180\,\text{g}\,C_6H_{12}O_6} \times \frac{2\,\text{mol}\,C_2H_6O}{1\,\text{mol}\,C_6H_{12}O_6} \times \frac{46.0\,\text{g}\,C_2H_6O}{1\,\text{mol}\,C_2H_6O} = 128\,\text{g}\,C_2H_6O$$

3.   a.
$$2Al + 6NaOH \;\rightarrow\; 2Na_3AlO_3 + 3H_2$$

$$2.5\,\text{mol Al} \times \frac{3\,\text{mol H}_2}{2\,\text{mol Al}} = 3.75\,\text{mol H}_2$$

   b.
$$2Al + 6NaOH \;\rightarrow\; 2Na_3AlO_3 + 3H_2$$

The molar mass of NaOH is 40.0 g.

$$56\,\text{g Al} \times \frac{1\,\text{mol Al}}{27.0\,\text{g Al}} \times \frac{6\,\text{mol NaOH}}{2\,\text{mol Al}} \times \frac{40.0\,\text{g NaOH}}{1\,\text{mol NaOH}} = 249\,\text{g NaOH}$$

Chapter 14 Practice Problems

1. A balloon has a volume of 4.0 L at 1 atm. What is the volume of the balloon if the pressure is decreased to 0.50 atm? (Hint: use Boyle's law)

2. A piston with a volume of 150.0 mL and a pressure of 742 torr is compressed to 50.0 mL. What is the new pressure? (Hint: use Boyle's law)

3. A balloon has a volume of 2.5 L at 22 °C. What is its volume at 205 °C ? (Hint: use Charles' law)

4. A gas is heated in a constant volume from 25 °C to 121 °C. If the pressure is initially 1.0 atm, what is the new pressure? (Hint: use Gay-Lussac's law)

5. A gas sample has a volume of 1.5 L at STP (0 °C and 1 atm). The temperature is increased to 200 °C and the volume increased to 2.2 L. What is the new pressure of the gas? (Hint: use the combined gas law)

6. What is the volume of 2.0 moles of nitrogen gas at 0 °C and 1 atm? (Hint: use the ideal gas law)

Solutions to Chapter 14 Problems

1. $P_1 = 1\,\text{atm}$ $\qquad\qquad P_2 = 0.50\,\text{atm}$

    $V_1 = 4.0\,\text{L}$ $\qquad\qquad V_2 = ?$

    $P_1V_1 = P_2V_2$ $\qquad V_2 = V_1 \times \dfrac{P_1}{P_2} = 4.0\,\text{L} \times \dfrac{1.0\,\cancel{\text{atm}}}{0.50\,\cancel{\text{atm}}} = 8.0\,\text{L}$

2. $P_1 = 742\,\text{torr}$ $\qquad P_2 = ?$

    $V_1 = 150.0\,\text{mL}$ $\qquad V_2 = 50.0\,\text{mL}$

    $P_1V_1 = P_2V_2$

    $P_2 = P_1 \times \dfrac{V_1}{V_2} = 742\,\text{torr} \times \dfrac{150.0\,\cancel{\text{mL}}}{50.0\,\cancel{\text{mL}}} = 2230\,\text{torr} = 2.23 \times 10^3\,\text{torr}$

3. $V_1 = 2.5\,\text{L}$ $\qquad\qquad\qquad V_2 = ?$

    $T_1 = 22 + 273 = 295\,\text{K}$ $\qquad T_2 = 205 + 273 = 478\,\text{K}$

    $\dfrac{V_1}{T_1} = \dfrac{V_2}{T_2}$ $\qquad V_2 = V_1 \times \dfrac{T_2}{T_1} = 2.5\,\text{L} \times \dfrac{478\,\cancel{\text{K}}}{295\,\cancel{\text{K}}} = 4.0\,\text{L}$

4. $P_1 = 1.0\,\text{atm}$ $\qquad\qquad\qquad P_2 = ?$

    $T_1 = 25 + 273 = 298\,\text{K}$ $\qquad T_2 = 121 + 273 = 394\,\text{K}$

    $\dfrac{P_1}{T_1} = \dfrac{P_2}{T_2}$ $\qquad P_2 = P_1 \times \dfrac{T_2}{T_1} = 1.0\,\text{atm} \times \dfrac{394\,\cancel{\text{K}}}{298\,\cancel{\text{K}}} = 1.3\,\text{atm}$

5. $P_1 = 1\,\text{atm}$ $\qquad\qquad\qquad P_2 = ?$

    $V_1 = 1.5\,\text{L}$ $\qquad\qquad\qquad V_2 = 2.2\,\text{L}$

    $T_1 = 273\,\text{K}$ $\qquad\qquad\qquad T_2 = 200 + 273 = 473\,\text{K}$

    $\dfrac{P_1V_1}{T_1} = \dfrac{P_2V_2}{T_2}$ $\qquad P_2 = \dfrac{P_1V_1T_2}{V_2T_1} = P_1 \times \dfrac{V_1}{V_2} \times \dfrac{T_2}{T_1}$

    $P_2 = 1\,\text{atm} \times \dfrac{1.5\,\cancel{\text{L}}}{2.2\,\cancel{\text{L}}} \times \dfrac{473\,\cancel{\text{K}}}{273\,\cancel{\text{K}}} = 1.2\,\text{atm}$

6. $P = 1.00\,\text{atm}$ $\quad n = 2.00\,\text{mole}$ $\quad V = ?$ $\quad T = 273\,\text{K}$

    $PV = nRT$

    $V = \dfrac{nRT}{P} = \dfrac{2.00\,\cancel{\text{mol}} \times 0.0821\,(\text{L}\,\text{atm}/\text{K}\,\cancel{\text{mol}}) \times 273\,\text{K}}{1.00\,\cancel{\text{atm}}} = 44.8\,\text{L}$

Chapter 15 Practice Problems

1. How many grams of glucose are there in 300 mL of 5.0% (w/v) solution?

2. If a patient receives 45.0 mL of 0.90% (w/v) NaCl intravenous solution, how many grams of NaCl does she get?

3. How many mL of 2.0% (w/v) KCl do you need to pour out in order to have 5.0 grams of KCl?

4. How many milliliters of alcohol are there in 472 mL (about 8 fl. oz.) of wine that has a 15% (v/v) alcohol content?

5. What is the molarity of a solution if 0.25 moles of $NaHCO_3$ is dissolved in water to make a total of 500.0 mL?

6. How many moles are there in 50.0 mL of 0.10 M HCl?

7. How many grams are there in 250.0 mL of 0.875 M K_2HPO_4? The molar mass of K_2HPO_4 is 174.2 grams/mol.

8. How many grams of K_2HPO_4 do you need to weigh out to make 250.0 mL of a 0.875 M solution?

9. How many milliliters of 2.0 M HCl do you need to measure out to obtain 0.50 moles of HCl?

Solutions to Chapter 15 Problems

1. $300 \, \text{mL} \times \dfrac{5.0 \, \text{g glucose}}{100 \, \text{mL}} = 15 \, \text{g glucose}$

2. $45.0 \, \text{mL} \times \dfrac{0.90 \, \text{g NaCl}}{100 \, \text{mL}} = 0.41 \, \text{g NaCl}$

3. $5.0 \, \text{g KCl} \times \dfrac{100 \, \text{mL}}{2.0 \, \text{g KCl}} = 250 \, \text{mL}$

4. $472 \, \text{mL} \times \dfrac{15 \, \text{mL alcohol}}{100 \, \text{mL wine}} = 71 \, \text{mL alcohol}$

5. $500.0 \, \text{mL} = 0.500 \, \text{L}$

   $\dfrac{0.25 \, \text{mol NaHCO}_3}{0.500 \, \text{L}} = \dfrac{0.50 \, \text{mol NaHCO}_3}{1 \, \text{L}} = 0.50 \, \text{M NaHCO}_3$

6. $50.0 \, \text{mL} \times \dfrac{1 \, \text{L}}{1000 \, \text{mL}} \times \dfrac{0.10 \, \text{mol HCl}}{1 \, \text{L}} = 0.0050 \, \text{mol HCl}$

7. $250.0 \, \text{mL} \times \dfrac{1 \, \text{L}}{1000 \, \text{mL}} \times \dfrac{0.875 \, \text{mol}}{1 \, \text{L}} \times \dfrac{174.2 \, \text{g K}_2\text{HPO}_4}{1 \, \text{mol}} = 38.1 \, \text{g K}_2\text{HPO}_4$

8. $38.1 \, \text{g K}_2\text{HPO}_4$

   This problem is exactly the same as the previous problem. Problem 7 asks how many grams are in 250 mL of solution. This problem asks how many grams do you put into 250.0 mL to make a 0.875 M solution.

9. $0.50 \, \text{mol HCl} \times \dfrac{1 \, \text{L}}{2.0 \, \text{mol HCl}} \times \dfrac{1000 \, \text{mL}}{1 \, \text{L}} = 250 \, \text{mL}$

Chapter 16 Practice Problems

1. In a water solution, $[H_3O^+]$ is 1.0×10^{-3} M. What is $[OH^-]$?

2. In another solution, $[OH^-]$ is 4.45×10^{-5} M. What is $[H_3O^+]$?

3. Calculate the pH for each of the following solutions:

    a. $[H_3O^+] = 4.50\times10^{-8}$ M

    b. $[H_3O^+] = 2.44\times10^{-3}$ M

    c. $[H_3O^+] = 7.90\times10^{-11}$ M

    d. $[H_3O^+] = 1.0\times10^{-5}$ M

4. For the following pH values, calculate $[H_3O^+]$

    a. pH = 1.3

    b. pH = 9.3

    c. pH = 6

    d. pH = 13

Solutions to Chapter 16 Problems

1. $K_w = [H_3O^+][OH^-]$

    $$[OH^-] = \frac{K_w}{[H_3O^+]} = \frac{1.00\times10^{-14}}{1.0\times10^{-3}} = 1.0\times10^{-11}\,M$$

2. $K_w = [H_3O^+][OH^-]$

    $$[H_3O^+] = \frac{K_w}{[OH^-]} = \frac{1.00\times10^{-14}}{4.45\times10^{-5}} = 2.25\times10^{-10}\,M$$

3. a.   7.35

    4.5 (EXP) 8 (+/-) (log) (+/-)   Display reads   7.34678

    b.   2.61

    2.44 (EXP) 3 (+/-) (log) (+/-)   Display reads   2.61261077

    c.   10.1

    7.9 (EXP) 11 (+/-) (log) (+/-)   Display reads   10.1023729

    d.   5   You don't need a calculator for this problem.

4. a.   0.050 M or $5.0\times10^{-2}\,M$

    1.3 (+/-) (INV) (log)   Display reads   0.0501187 or 5.01187 -02

    Remember, your calculator may have a shift or 2nd button instead of (INV).

    b. $5.0\times10^{-10}\,M$

    9.3 (+/-) (INV) (log)   Display reads   5.0118723 -10

    c. $1.0\times10^{-6}\,M$ Since the coefficient is 1, we don't need to use a calculator.

    6 (+/-) (INV) (log)   Display reads   0.000001 or 1.000000 -06

    d. $1.0\times10^{-13}\,M$ Since the coefficient is 1, we don't need to use a calculator.

    13 (+/-) (INV) (log)   Display reads   1.0000000 -13